ESPRIT Basic Research Series

Edited in cooperation with
the Commission of the European Communities, DG XIII

R. Kassing (Ed.)

Scanning Microscopy

Symposium Proceedings

Springer-Verlag
Berlin Heidelberg New York
London Paris Tokyo
Hong Kong Barcelona
Budapest

Volume Editor

Rainer Kassing
Institut für Technische Physik, Universität Kassel
Heinrich-Plett-Str.40, W-3500 Kassel, FRG

ISBN-13:978-3-642-84812-4 e-ISBN-13:978-3-642-84810-0
DOI: 10.1007/978-3-642-84810-0

Library of Congress Cataloging-in-Publication Data
Scanning microscopy : symposium proceedings / R. Kassing (ed.). p. cm. --
(ESPRIT basic research series) Includes bibliographical references.
ISBN-13:978-3-642-84812-4

1. Scanning tunneling microscopy--Congresses. I. Kassing, R. (Rainer)
II. Series. QH212.S35S25 1992 502'.8'25--dc20 92-24849

Publication No. EUR 14433 EN of the Commission of the European Communities, Scientific and Technical Communication Unit, Directorate-General Telecommunications, Information Industries and Innovation, Luxembourg

Softcover reprint of the hardcover 1st edititon 1992

Typesetting: Camera ready by authors
45/3140 - 5 4 3 2 1 0 - Printed on acid-free paper

Foreword

With the invention of the scanning tunneling microscope in 1982 by Binnig and Rohrer and the subsequent award of the Nobel Prize, the field of scanning microscopy was given a strong boost in view of its wide range of applications.

In particular, expanding the capability to access nature's foundations at the atomic level is now recognized as having the potential for major impact in Information Technology.

This third volume of the ESPRIT Basic Research Series provides a well structured overview of the state of the art of scanning microscopy and recent advances including results of ESPRIT Basic Research Actions 3109 and 3314.

April 1992 G. Metakides

Preface

The IMO Symposium Fall '90, Wetzlar, FRG, October 1/2, 1990, brought together leading scientists and researchers in scanning microscopy from research institutes and industries, each of whom was invited to contribute a lecture which was followed by a discussion. The resulting contributions are contained in this proceedings.

Microscopic techniques are used not only for research work in material and life science but also for routine applications in almost any vital section of our everyday life. The demand for coming to a better understanding of materials and their behaviour under different conditions and environments as well as all aspects of human life initiated an ongoing development for improved microscopic techniques.

The history of scanning microscopy goes back to the early 1930s when M. von Ardenne developed the first scanning electron microscope immediately after the fundamental work of E. Ruska and co-workers had been carried out in electron optics and electron microscopy. From this early work it took more than 30 years until the first commercial version of a scanning electron microscope became available to the scientific community. Since then scanning electron microscopy has become a well known technique used not only for research in a wide field of different applications but also for routine quality control and failure analysis. The Programme Committee of the IMO Symposium Fall '90 decided to complement the scheme of scanning microscopy with recent developments in scanning optical and acoustic microscopy and especially the very fast growing field of probe techniques.

More specifically, in the 1950s M. Minsky got a patent for scanning optical microscopy but it was not until 1984 that the first commercial instrument for material science applications became available. But not before the development of a system dedicated specifically to fluorescence applications had been finished the instrument became more widely spread. Today most of the instruments are used either for critical dimensions measurements or for applications in life science.

The development of the scanning acoustic microscope goes back to a patent of Sokolov in the 1940s but due to technical reasons the first high-resolution scanning acoustic microscope was developed by C.F. Quate in 1978. Since 1985 instruments have been commercially available especially for applications in the semiconductor industry, for defect control and for applications in material science.

In the early 1980s G. Binnig, H. Rohrer, and co-workers from the IBM Research Laboratory Zürich invented a new technique of scanning microscopy for high-resolution characterization of surfaces. For their work resulting in the scanning tunneling microscope (STM) Binnig and Rohrer shared the 1986 Nobel Prize in Physics with E. Ruska for his work in electron microscopy. After the first results on scanning tunneling microscopy were published showing the potential of this new technique, a large number of laboratories all over the world have developed their own STMs for specific applications and have started to develop new techniques utilizing different distance-dependent signals instead of the tunneling current to characterize surfaces under different aspects. Within a few years STMs became also commercially available for a range of applications.

Therefore the Programme Committee compiled high quality lectures on the techniques of scanning microscopy as well as of current applications in order to provide to the audience a complete overview of the state of the art.

April 1992

The Programme Committee
IMO Symposium Micro-System Technology

Contents

Performance and selection criteria of critical components of STM and AFM

Yves Martin
IBM T.J. Watson Research Center
P.O.Box 218
Yorktown Heights, N.Y.10598

Abstract

This paper describes the working principles of STM, and of AFM working in both repulsive and attractive force modes, and addresses the implementation and performance criteria of some of the critical components of these instruments. The emphasis is the design of instruments suitable for technological applications. The paper examines criteria for the tips, scanners, and general optical and mechanical configurations. An example is given for a solution that we adopted.

Introduction

Since their invention in 1982 and 1986 respectively {1,2}, the scanning tunneling microscope (STM) and the atomic force microscope (AFM) have served very fruitful basic research purposes. They have enabled the observation and study of the chemistry and organisation of crystal surfaces, in many cases down to the atomic level. This paper examines the implementation and performance criteria of several critical components of the STM and AFM, particularly those with a strong impact on technological applications.

Technological applications for STM and AFM in the microelectronic and optical industry are growing in importance. Engineers have realized that STM and AFM distinguish themselves from traditional forms of microscopies in many aspects, and can significantly increase the available range of various measurements. They operate in a near-field configuration, they have capabilities to detect and to control motions down to the angstrom level. Figure 1 demonstrates the unique capabilities of the SFM and AFM for imaging and measuring dimensions of small sample features on a nanometer scale. SFM and AFM are compared to several other types of microscopies, with regard to their lateral and vertical imaging ranges. They are the only instruments to fill the lower left part of the diagram, that is to have simultaneous capabilities for vertical and lateral nanometer resolution.

Operating principles

In both STM {3} and AFM {3,4}, a very sharp tip is positioned at close proximity to an object and moved in a raster fashion accross the object. Both techniques measure some type of close interaction between the tip and the sample, in order to form an image and also in order to adjust the tip-sample spacing to a very small value in an automatic fashion.

STM requires the sample to be electrically conducting. The metallic tip is biased to a small voltage, usually a fraction of a volt. When the tip is within 10 Å from the sample a small current can flow between the tip and the sample, usually of order of one nanoampere. The underlying physical principle for this current is vacuum tunneling of electrons {6}. One key property of this principle tells us that the current depends strongly on the tip to sample spacing. The tunneling current decreases by roughly an order of magnitude for every distance increase by 1 Å, thus giving the extreme sensitivity to changes in sample height. It also results in the confinement of the tunneling current to the most protruding atom of the tip, which in turn, is the key to atomic resolution.

The AFM resembles a very sensitive and gentle stylus profilometer {7}. The stylus or tip is carried by a cantilever beam, and the interacting quantity is the force between the tip and the sample. Two modes of operation are identified, where different principles are applied, and which are implemented in different forms:

1. The interacting quantity is primarily the repulsive force between the last atom of the tip and the nearest atom(s) of the sample {8}. In this mode, most similar to the stylus profilometer, tip and sample are in close contact. The tip traces the sample surface with a constant loading force. The deflection of the cantilever is measured and automatically readjusted to be kept constant, in a feed-back manner. The key for high sensitivity lies in the choice of the detector for the vertical motion of the tip. Initially, Binnig et al {2} chose to use a second tip, maintained a few Å away from the back of the cantilever, and operated as a tunneling tip. Later, laser-based motion detectors have served the same function. In the best operating conditions, the minute vertical motion of the cantilever due to the sample atomic corrugation is observed; atomic resolution is obtained by the AFM, when operating in this mode.
2. The second mode {9} relies on the detection of long range forces between the tip and cantilever. Among those are the van der Waals forces and dipolar forces {10}, electrostatic forces, and magnetic forces. These forces are still detectable when the tip is separated from the sample by a few tens of Å. The key to detection sensitivity and stability is to resort to some form of A.C. measurement, instead of directly measuring the deflection induced by the force. For this purpose, the cantilever, mounted on a piezoelectric holder, is vibrated by a minute amount (a few Å) at a frequency close to its resonance. The vertical gradient of the tip-sample force shifts the resonance. This small resonance shift leads to a much larger amplitude change, because of the mechanical quality factor Q of the cantilever (a typical value of Q is 100). The detected amplitude serves as a feed-back signal to keep the force gradient constant. In most practical situations, this also keeps the spacing between tip and sample constant. Hence, the tip closely follows the sample topography. Lateral spatial resolution in this mode can approach 10 Å, and vertical resolution can be smaller than one Å.

The overall choice between STM and AFM depends on the type of sample to examine and on the range of applications to be satisfied. STM can only inspect conducting samples, or samples that can be coated with a conducting layer (such as gold). AFM can image conductors and insulators, but they also are more complex instruments. Our own efforts have centered on the second mode of AFM (see above) also being referred to as the attractive force type AFM, or as the laser force microscope {5}. Its function is general non-contact, non-destructive examination and profiling. It also has additional capabilities to map magnetic field {11} and perform electrical measurements on a scale of the order of a few tens of nanometers {12}.

Key components of the instruments

Figure 2 shows the key components for an STM or AFM, conceived as a general measurement tool for technological applications. The choice and quality of these components will determine the performances of the instrument, and the capacity for automation of instrument control and of acquisition of data. They include the tip, the scanner, the approach mechanism, the positioning mechanism, the tip motion sensor for the AFM, and the optical viewer used for sample positioning.

Tips for STM can take the most simple form, such as a clipped or etched tungsten wire. Noble metals are also chosen for their oxide-free surface and for stiffness {13}. When inspecting very flat surfaces, the details of the shape are often unimportant, apart from the very end. For many technological applications however, this is not true. When mapping microroughness, the sharpness of the tip end limits the resolving power {14}. Methods for fabricating sharp tips abound in the literature, partly because of their past utilization in electron and field ion microscopes {15}. On the other hand, for metrology applications, the shank of the tip plays a key part when locating edges and measuring line-widths, and recent articles have described solutions {16}.

The AFM tip is a more elaborate component, since the tip needs to be mounted on (or be part of) a cantilever, and since all dimensions of this component lay in the micron range. Fabrication of this key component has evolved from "artisan" construction methods, to more reproducible lithographic processes, based on silicon micro-machining technology (see figure 3). In addition, several new ways to control the tip shape and sharpness have been the theme of a few recent efforts {4,17}.

Scanners {18} in STM and AFM have unanimously made use of piezo-actuators for their active element. They are usually adequate for the small required range of motion. Their advantages include a high stiffness, and hence high resonance frequencies and high speed scanning capabilities, low thermal dissipation from the electrical drive and a high thermal stability, and simplicity of construction. The pioneer the piezo-tripod, was followed by the piezo tube scanner, and by various flexure schemes {19,20}. The piezo-element however exhibits the disadvantage of high non-linearity, due to hysteresis and creep. Accuracy can nontheless be attained through the use of independent calibrated motion sensors, such as heterodyne interferometers, or strain or capacitance sensors {20}.

The mechanical assembly that holds the scanner and tip or sample needs to provide for prepositioning, viewing, and gentle but precise approach of tip to sample. Original solutions to the approach and positioning problem are found in the "louse" actuator {1,21} or in the inertial "walker" {22}. Of high interest also is the NIST molecular measuring machine project {23}. This design centers on accessing a large measurement field, of 50 mm x 50 mm x 0.1 mm, simultaneously with a very high resolution of 0.1 nm and absolute an accuracy of 1 nm. These high risk goals put the most stringent demands on the accuracy and stability of the mechanisms and of the motion sensors.

The AFM necessitates an additional key component, a highly sensitive tip motion sensor. Several laser-based sensors have been adopted for this task. The simplest is the laser beam deflection sensor {24}. Most other sensors are interferometrical sensors, that operate either on homodyne or heterodyne principles {9}. They are built out of discrete bulk optical components or optical fibers, and use gas or solid-state lasers {25-28}. Capacitive sensors have also been adopted {29}.

The choice of a particular sensor for a specific AFM instrument is an important but complex issue. Some sensors that can prove adequate for a research instrument may not be suitable for a user-friendly instrument. Among the selection criteria, we find sensitivity (or smallest detectable motion), range of detectable motion (largest measurable motion, d.c. or a.c. motion, and frequency range), amount of induced perturbation on the tip, simplicity of tuning and operation, and many practical implementation criteria such as compatibility with scanning, with prepositioning and approach mechanisms, or with optical viewing.

Table 1 compares some of these implementation criteria for a laser heterodyne interferometer sensor {25}, and for a laser beam deflection sensor {24}. The advantages of the deflection technique are points 6 and 9: ability to measure D.C. displacements, and simplicity. For this last point however, the difference is small: The deflection sensor could become fairly complex if scanning the tip is a requirement. We chose the heterodyne detection for our instrument because of our requirement to scan the tip, and also partly because of our past experience with this technology. We intended to inspect samples larger than one hundred millimeters that cannot easily be scanned in a rapid and precise motion. Figure 4 shows the basic configuration for the optics and the tip scanning geometry. Both the objective lens and the tip are moved by the piezo scanner: in this manner, the laser beam of the heterodyne interferometer stays focused on the back of the tip cantilever.

The main application of STM and AFM is imaging of structures on a sub-micron to nanometer scale, where optical microscopy is not usable. Figure 5 shows pictures of a test sample, V-grooves etched in silicon <100>, taken with our attractive force type AFM. Accurate dimensions of these grooves can be extracted from these pictures, along the horizontal and vertical directions. In addition, the $1\mu m \times 1\mu m$ enlargment can be used to measure the surface roughness of the sample: One can notice that the top flat surface is smoother than the sloped part of the groove, which is filled with small terraces a few 100 Å in size.

Conclusion

The design of an STM or AFM instrument and the choice of its critical components depends on the type of applications the instrument ought to perform. For general inspection and dimensional measurement in the microelectronic industry, we chose the attractive force type AFM, which works in a non-contact mode. Choices of tips, scanner, and optical sensor was influenced both by performance requirements and flexibility of operation. This instrument will undoubtedly become a valuable tool to accompany -or in some cases replace- an SEM, for inspection in the sub-micron domain.

References

1. G.Binnig and H.Rohrer, "Scanning tunneling microscopy", Helv.Phys.Acta 55, 726 (1982)
2. G.Binnig, C.F.Quate, and Ch.Gerber, "Atomic force microscope", Phys.Rev.Lett.56, 930 (1986)
3. review papers: G.Binnig and H.Rohrer, "The scanning tunneling microscope", Scientific American 253, 50 (1985) or "Scanning tunneling microscopy", IBM J.Res.Develop. 30, 355 (1986)
4. P.K.Hansma, V.B.Elings, O.Marti, C.E.Bracker, "Scanning tunneling microscopy and atomic force microscopy: application to biology and technolgy", Science 242, 157 (1988)
5. Y.Martin, C.C.Williams, and H.K.Wickramasinghe, "Tip-techniques for microcharacterization of materials", Scanning Microscopy 2, 3 (1988)
   H.K.Wickramasinghe, "Scanned-probe microscopes", Scientific American 260, 98 (1989)
6. C.G.Duke, "Tunneling in solids", Academic Press, New-York, 1969
   J.Simmons, "Generalized formlua for the electric tunnel effect between similar electrodes separated by a thin insulating film", J.Appl.Phys.34 1793 (1963)

P.Hansma, "Tunneling spectroscopy: capabilities, applications and new techniques", Plenum Press New-York, 1982

7. J.B.P.Williamson, "Microphotography of surfaces", Proc.Inst.Mech.Eng.London 182, 21 (1967-68)
M.Stedman and K.Lindsey, "Limits of surface measurements by stylus instruments", SPIE Surface measurement and characterization 1009, 56 (1988)

8. A.L.Weisenhorn, P.K.Hansma, T.R.Albrecht, and C.F.Quate, "Forces in atomic force microscopy in air and water", Appl.Phys.Lett.54, 2651, (1989)
J.B.Pethica, and W.C.Oliver, "Tip surface interactions in STM and AFM", Physica Scripta T19, 61 (1987)

9. G.McClelland, R.Erlandson and S.Chiang, "Atomic force microscopy: general principles and a new implementation", Rev.Prog.in Quantitative Non-Destructive Evaluation 6B, ed by D.O Thompson and D.E.Chimenti, 1307 (1987)
Y.Martin, C.C.Williams, and H.K.Wickramasinghe, "Atomic force microscope - force mapping and profiling on a sub 100-Å scale", J.Appl.Phys. 61, 4723 (1987)

10. J.N.Israelachvili, "Intermolecular and surface forces", Academic Press London, 1985

11. Y.Martin, D.W.Abraham, P.C.D.Hobbs, and H.K.Wickramasinghe, "Magnetic force microscopy - a short review", Electrochemical Soc. Proc. on Magnet.Mat.Proces.Dev. 90-8, 115 (1989)
D.Rugar, H.J.Mamin, P.Guethner, S.E.Lambert, J.E.Stern, I.McFadyen, T.Yogi, "Magnetic force microscopy: general principles and application to longitudinal recording media", to be published.
C.Schonenberger and S.F.Alvarado, "Understanding magnetic force microscopy", to be published.

12. Y.Martin, D.W.Abraham, and H.K.Wickramasinghe, "High-resolution capacitance measurement and potentiometry by force microscopy", Appl.Phys.Lett. 52, 1103 (1988)

13. U.Ducrig and O.Zueger and D.W.Pohl, "Force sensing in scanning tunneling microscopy: observation of adhesion forces on clean metal surfaces", J. of Microscopy 152, 259 (1988)

14. J.K.Gimzewski and A.Humbert, "STM of surface microstructure on rough surfaces", IBM J.Res.Develop. 30, 472 (1986)

15. F.W.Niemeck and D.Ruppin, "Electro-etching of cathode tips for field-electron microscopes", Z.Angew.Phys. 6, 1 (1954)
H.W.Fink, "Mono-atomic tips for STM", IBM J.Res.Develop. 30, 460 (1986)
Vu Thien Binh, "In situ fabrication and regeneration of microtips for STM", J. of Microscopy 152, Pt 2, 355 (1988)
D.K.Biegelsen, F.A.Ponce and J.C.Tramontana, "Simple ion milling preparation of < 111 > tungsten tips", Appl.Phys.Lett.54, 1223 (1989)

16. Y.Akama, E.Nishimura, A.Sakai, H.Murakami, "New scanning tunneling microscopy tip for measuring surface topography", J. Vac. Sci. Technol. A, Vac. Surf. Films (USA) Vol.8, 429 (1990)

17. T.R.Albrecht and C.F.Quate, "Atomic resolution with the atomic force microscope on conductors and nonconductors", J.Vac.Sci.Technol.A 6,
R.B.Marcus, T.S.Ravi, T.Gmitter, K.Chin, D.Liu, W.J.Orvis, D.R.Ciarlo, C.E.Hunt, J.Trujillo, "Formation of silicon tips with < 1 nm radius", Appl. Phys. Lett. 56, 236 (1990)

18. D.W.Pohl, "Some design criteria in scanning tunneling microscopy", IBM J.Res.Develop. 30, 417 (1986)

19. G.Binnig and D.Smith, "Single tube three dimensional scanner for scanning tunneling microscopy", Rev.Sci.Instrum. 57, 1688 (1986)
R.G.Carr, "Finite element analysis of PZT tube scanner motion for STM", J.of Microscopy 152, Pt2, 379 (1988)

P.Muralt, D.W.Pohl, W.Denk, "Wide low-range low-operating voltage bimorph STM: application as potentiometer", IBM J.Res.Develop. 30, 443 (1986)
F.S.Scire and E.C.Teague, "Piezodriven 50-μm range stage with subnanometer resolution", Rev.Sci.Instr. 49, 12 (1978)
A novel suggestion for miniature scanner is:
T.Albrecht, S.Akamine, M.J.Zdeblick, and C.F.Quate, "Microfabrication of integrated STM", J.Vac.Sci.Technol. A8, 317 (1990)

20. A.Stemmer, A.Engel, R.Haring, R.Reichelt and U.Aebi, "Miniature-size STM with integrated 2-axes heterodyne interferometer and light microscope", Ultramicroscopy 25, 171 (1988)
H.Yamada, T.Fujii and K.Nakayama, "Linewidth measurement by a new STM", Jap.J.Appl.Phys. 28, 2402 (1989)
N.K.Reay, P.D.Atherton and T.R.Hicks, "STM: absolute positionning using capacitance micrometry"
P.Atherton, "Nano positioning and control", Electro Opt. (UK) 18, 15 (1988)

21. Ch.Gerber, G.Binnig, H.Fuchs, O.Marti and H.Rohrer, "STM combined with a scanning electron microscope", Rev.Sci.Instrum. 67, 221 (1986)

22. K.H.Besocke, "An easily operable STM", Surf.Sci. 181, 145 (1987)

23. E.C.Teague, "The National Institute of Standards and Technology molecular measuring machine project: metrology and precision engineering design", J.Vac.Sci.Technol. B7, 1898 (1989)
Simpler but more practical designs can be found in:
B.L.Blackford, D.C.Dahn, and M.H.Jericho, "High-stability bimorph STM", Rev.Sci.Instrum. 58, 1343 (1987)
O.Marti, B.Drake and P.K.Hansma, "AFM of liquid-covered surfaces: atomic resolution images", Appl.Phys.Lett. 51, 484 (1987)
M.Yasutake and C.Miyata, "STM combined with optical microscope for large sample measurement", J.Vac.Sci.Technol. A8, 350 (1990)

24. G.Meyer and N.M.Amer, "Novel optical approach to AFM", Appl,Phys.Lett. 53, 1045 (1988)

25. D.Royer, E.Dieulesaint, and Y.Martin, "Improved version of a polarized beam heterodyne interferometer", Proc. of IEEE Ultrasonics Symp. S.Francisco, ed. by B.R.McAvoy (IEEE, New York, 1985), p.432

26. C.Schonenberger and S.F.Alvarado : "A differential interferometer for force microscopy", Rev.Sci.Instrum. 60. 3131 (1989)

27. D.Rugar, H.J.Mamin, R.Erlandsson, J.E.Stern, and B.D.Terris, "Force microscope using a fiber-optic displacement sensor", Rev.Sci.Instrum. 59, 2337 (1988);

28. D.Sarid, D.A.Iams and V.Weissenberger, "Compact scanning force miscroscope using a laser diode", Opt.Lett.13, 1057 (1988)

29. T.Goddenhenrich, U.Lemke, U.Hartmann, and C.Heiden, "Force microscope with capacitive displacement detection", J.Vac.Sci.Technol. A8, 383 (1988)
G.Neubauer, S.R.Cohen, and G.M.McClelland, in Interfaces between Polymers, Metals, and Ceramics, ed by B.M.DeKoven, A.J.Gellman, and R.Rosenberg 153, (1989)

	heterodyne	**deflection**
1. Sensitivity	d_{min} = 5 10^{-3}Å for 1mW laser beam, BW = 10kHz, S/N = 1	d_{min} ×L/a L = cantilever length, a = spot diameter
2. Reflectivity of back of cantilever	Non critical: cantilever with tips or wire tips are usable	Only highly reflective cantilevers can be used
3. Approach and scanning the tip	Approach is parallel to the laser beam; tip and focusing lens are scanned	?? Can the tip be scanned? How much of the optics must also be scanned? With what accuracy?
4. Alignment of the laser beam on the cantilever	Using additional beam steering optics, and viewing the tip and laser beam with the optical viewer	By motion of the laser source or of the focusing lens
5. Optical viewing	from the top; the tip, the laser beam and the sample are seen; the magnification corresponds to 20x	?? Could be done from the top with the same focusing lens, but with a smaller magnification (5x). Alternative solution: using a second set of optics, and viewing from the side.
6. Measured displacement	A.C. displacement only	A.C. and D.C. displacement
7. Absolute measurement of displacement amplitude	Yes	Approximate
8. Sensitivity to thermal drift	Insensitive	Insensitive for a.c. measurement; ?? for d.c. measurement (a few 10 Å of drift over minutes?)
9. Simplicity of the design	Lots of optics (interferometer), can be packaged; a few special electronic components (HF receiver, limiter, phase demodulator)	Apparently simpler optics and electronics; However, what is the added complexity if scanning the tip is required?
10. Complexity of the adjustments	Automatic electronic adjustments; no optical adjustment when changing sample; limited optical adjustments when changing the tip (x,y and z focusing); some retuning of the interferometer every month or semester	?? Adjustments are probably simple, except if the tip is scanned; this is an unknown

Table 1. : Comparison of some performance and implementations criteria for a laser heterodyne interferometer sensor {25}, and for a laser beam deflection sensor {24}. Question marks correspond to issues that are still unclear or unknown to the author.

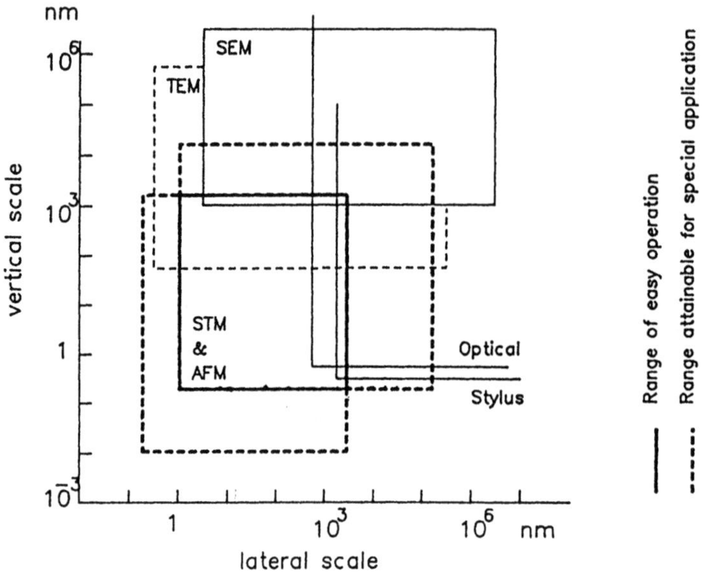

Figure 1: Lateral and vertical imaging ranges and spatial resolution of STM and AFM compared with those of other well established microscopes. STM and AFM provide the unique capability for simultaneous lateral and vertical nanometer resolution.

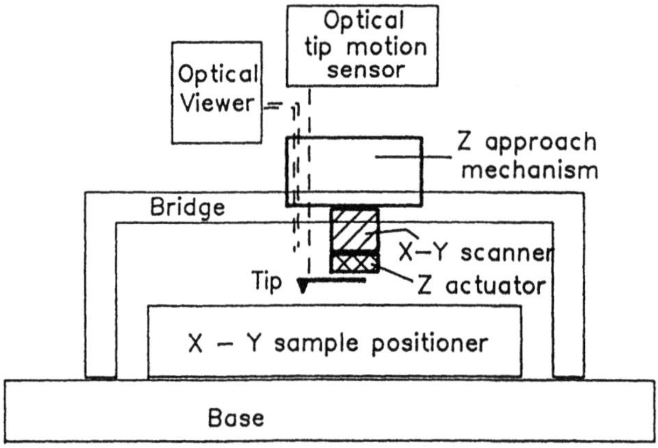

Figure 2: Block diagram of the mechanical and optical components for an AFM or STM. The configuration is for a general purpose tool, designed mainly for technological applications.

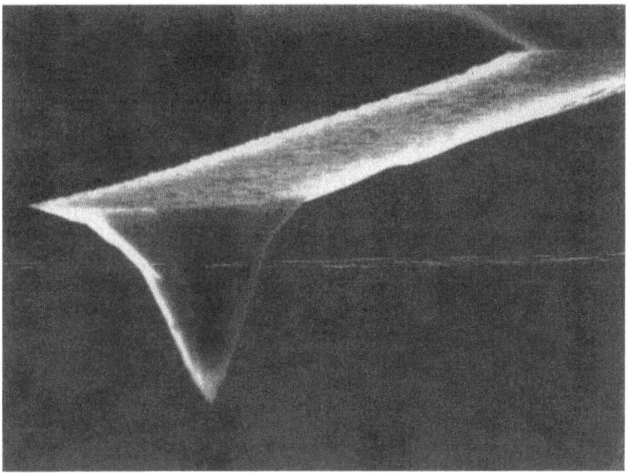

Figure 3: Silicon cantilever with integrated tip made by O.Wolter at IBM - GMTC Sindelfingen (W.Germany), using selective silicon etching techniques. Tips with heights between 2 and 30 micron have been fabricated.

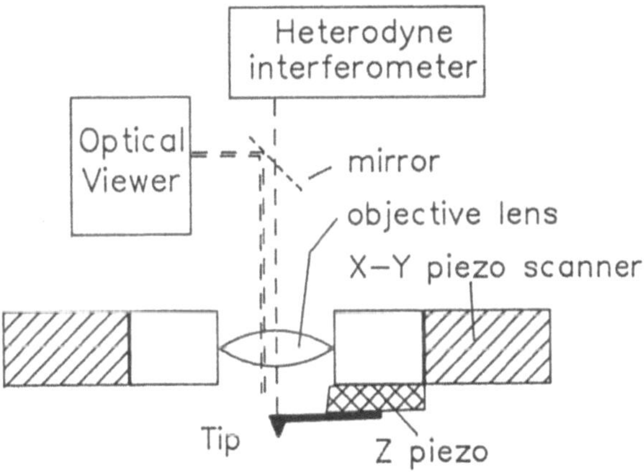

Figure 4: Configuration for tip, scanner and optics in our AFM working in the attractive force mode.

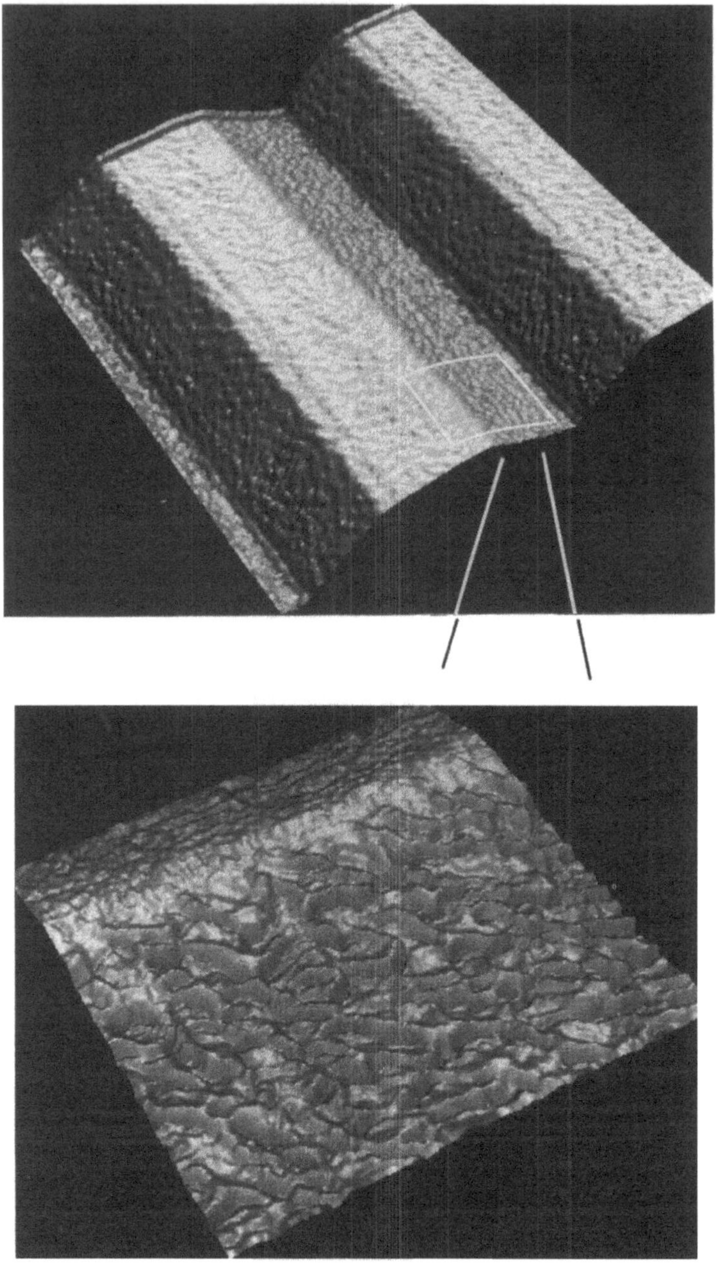

Figure 5: Example of images obtained with our AFM: V-grooves anisotropically etched in silicon <100>. Scan dimensions are $5 \mu m \times 5 \mu m$ for the top view, and $1 \mu m \times 1 \mu m$ for the bottom view.

Investigations on the SFM

- Tip to Substrate Interaction -

Prof. Dr. Rainer Kassing
Institute for Microstructuretechnology and Optoelectronics - IMO -, Wetzlar

Introduction

The scanning probe microscopy has been developed to a fascinating tool for surface characterisation. Because it is a nearfield method [1] high resolution can be obtained. Besides the Scanning Tunneling Microscope (STM) [2], optical [3], acoustical [4] and magnetic [5] nearfield microscopy, the Scanning Capacitance Microscope (SCM) [6], the Scanning Thermal Profiler [7] and the Scanning Ion Conductance Microscope [8], especially the Scanning Force Microscope (SFM) [9] has become of increasing interest due to its large field of applications.

Using this type of scanning probe microscopy a micro-tip scans the sample surface, and the tip-surface interaction is measured. There are two different configurations a) the static and b) the dynamic mode.

In the static mode the tip contacts the sample surface and the tip deviation is directly measured. The vertical forces are very small, in the range of $10^{-6} - 10^{-8}$ N, but several orders of magnitude larger than those in the dynamical mode. The distance dependence of the repulsive interaction forces between tip and surface is very large so that with a relatively simple apparatus a very high vertical and lateral resolution can be obtained. But due to the contact between tip und surface a high wear resistance of the tip is necessary. An advantage is, however, that it can be measured in nearly every medium.

In the dynamical mode the tip will be scanned on the surface in some distance without any contact to the sample surface. The typical distances are in the range from 2 nm to 20 nm, but they also can be in the range of μm. The tip surface interaction forces are several orders of magnitude smaller than in the static mode, and are in the range of $10^{-8} - 10^{-12}$ N. Because such small forces cannot be measured statically with the necessary sensitivity, not the forces itself but the force-gradient is determined by this method. For this reason the tip is placed on a cantilever and this system is stimulated to oscillate with its resonance frequency. The interaction between tip and surface is then measured as the change in resonance frequency, amplitude and/or Q-value. Usually the change in amplitude is measured. Since the distance dependence is not so strong in this method, an atomic lateral resolution is not possible. The lateral resolution lies in the order of the tip to surface distance. The disadvantage of the dynamic SFM is the lower lateral resolution due to the smaller tip to surface interaction, the advantage, however, is the nondestructive contactless scanning mode and furthermore new measuring techniques which are not possible in the static mode, e. g. the simultaneous measurement of the topography and capacitance, film thickness or dielectric constant as well as the possibility for high speed general mapping with scanning velocities up to 100 μm/s. In the following the different types of tip to surface interactions will be discussed. The used system allows scanning tunneling, - force, - capacitance, - acoustic and - potentiometric measurements [10, 11].

Experimental set-up of the SFM

Fig. 1 shows the block diagram, Fig. 2 the mechanical construction of the measuring head and Fig. 3 the principle diagram of the differential laser interferometer which is used to measure the amplitude of the vibrating cantilever. An electron micrograph of the used tips and cantilevers is shown in Fig. 4. A detailed description of these parts are given in [12]

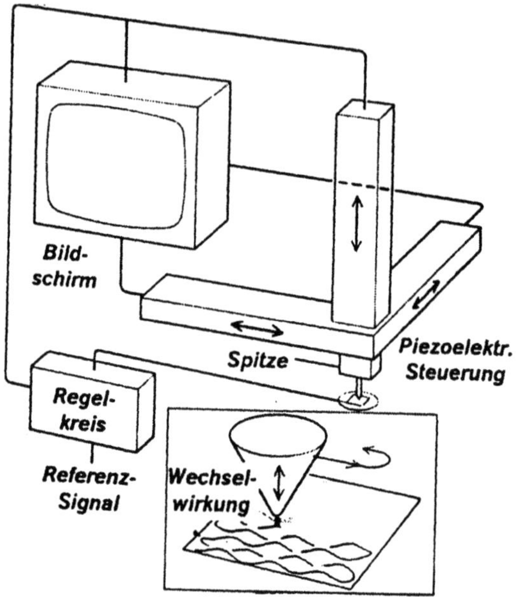

Fig. 1: Block diagram of the scanning probe microscopy

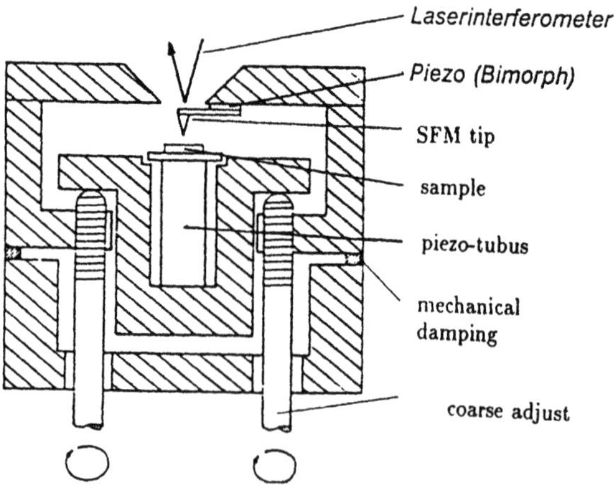

Fig. 2: Mechanical construction of the measuring head

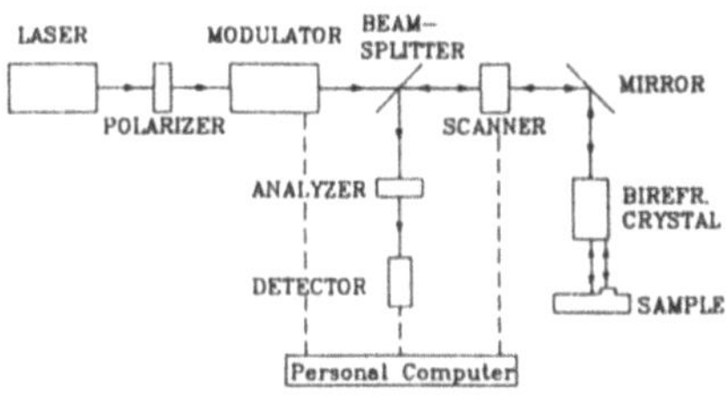

Interferometer set-up.

Fig. 3: Interferometer set-up

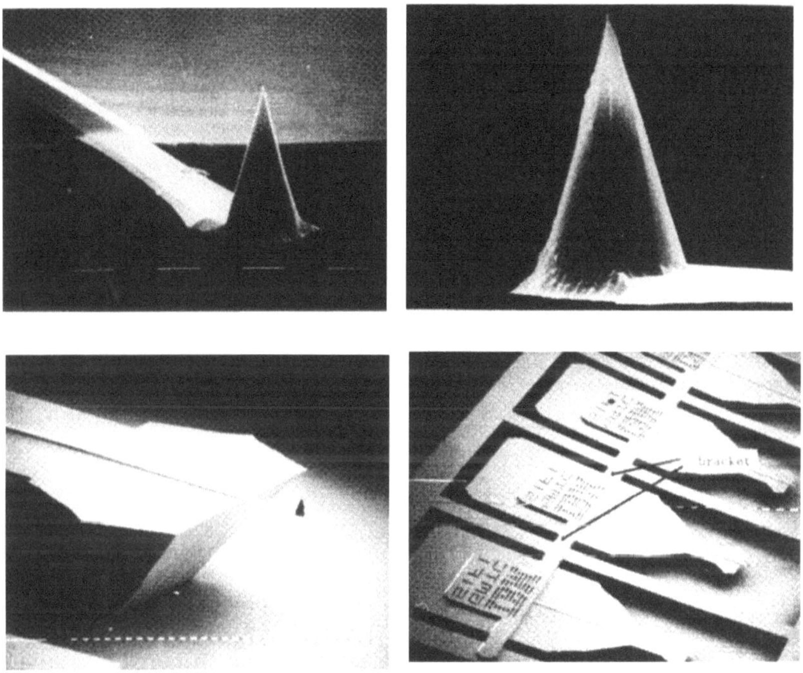

Fig. 4: SFM-tip prepared from silicon (by courtesy of J. Greschner and O. Wolter)

Properties of the tips and cantilevers

The characteristic parameters of the tips and cantilevers are the resonance frequency, force-constant, Q-value and tip radius of curvature and slope. If the frequency spectrum is known ν_{res} and Q-value can be determined and if length and width are known, k and thickness can be determined.

If the cantilever oscillates in gaseous on fluid medium the particles of these media will also oscillate together with the cantilever. Due to this energy loss a decreasing Q-value and due to the additional oscillating mass a decreasing resonance frequency results. The correct solution of the corresponding equations of oscillations is very complex. But because the Q-value in air is still relatively large (≥ 200), i. e. the damping factor is relatively small, it can be written

$$\nu_{res} = \nu_{res,0}\left(1 - \frac{1}{2Q^2}\right) \tag{1}$$

with $\nu_{res,0}$ the value of the resonance frequency for Q-value equal to infinity, i. e. without any damping. The change from $Q = \infty$ to $Q = 350$ for instance changes ν_{res} by a factor of 2×10^{-6}. Therfore the additional mass of the surrounding medium which oscillates together with the cantilever dominates the shift of the resonance frequency in this case and one obtains

$$\frac{\Delta\nu_{res}}{\nu_{res}} = -\frac{1}{2}\frac{\mu}{\rho A} \tag{2}$$

with μ= the corresponding additional mass per unit length and ρ the density and $A = b * d$ the product of width and thickness of the cantilever, i. e. ρ A is the mass of the cantilever per unit length.

For the amplitude of the forced oscillation one obtains the known Lorenz-profile:

$$A(\nu) = \frac{Qa_o \frac{\nu_{res}}{\nu}}{\sqrt{1 + Q^2\left(\frac{\nu_{res}}{\nu} - \frac{\nu}{\nu_{res}}\right)^2}} \tag{3}$$

with a_o the amplitude of the stimulating force or A/a_o the resonance step-up.

Experimental results

For the correct experimental determination of the interaction between tip and sample surface a more or less exact knowledge of the dependence of the resonance properties of tip and cantilever on the external parameters like temperature and pressure is necessary. For this reason measuring heads (tip plus cantilever) have been produced with force constants in the range of 0,01 N/m up to 1000 N/m with resonance-frequencies in the range of 1 kHz up to several MHz. In an optimal system the resonance frequency would be as large as possible, whereas for the Q-value and force-constant k a compromise adjusted to the type of interaction has to be found.

Fig. 5 shows the experimental result (dots) and the according to eq. (3) calculated resonance-curve for a Si cantilever with 3320 μm in length, 120 μm width an 23 μm thickness. The resonance frequency of 2.885 Hz and Q-value of 300 can be determined with high accuracy.

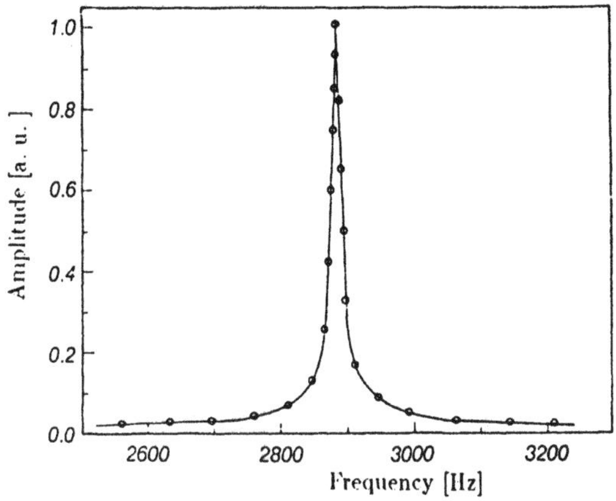

Fig. 5: Resonance curve of a Silicon cantilever

Pressure dependence

In air the Q-values of the cantilevers are in the range of 100 to 1000, whereas in UHV values of 10^5 are obtained, that means, as already mentioned that the additional mass of the air determines the damping. Therefore the Q-value must be pressure dependent. In a pressure range where the mean free path of the particles is small compared to the acustic wavelength of the oscillating cantilever, the air can be treated as an incompressible viscous fluid and one obtains with $\mu \simeq 2tb\rho_L$ (t = thicknes of the air film)

$$\frac{\Delta\nu}{\nu} = -\frac{\rho_L}{\rho_{Si}}\frac{t}{d} \tag{4}$$

Fig. 6 shows the experimental result in a linear as well as logarithmic scale for the Q-value and resonance frequency. The cantilever had the dimensions 1078 μm x 136 μm x 20 μm and the resonance frequency of 23,95 kHz. From Fig. 6 one determines with $\rho_L = 1,2 \text{kg/m}^3$, $\rho_{Si} = 2,33 * 10^3 kg/m^3$, t = 60 μm. Assuming that an air film also at the sidewalls of the cantilever exists, this values reduces to 40 μm. The acustic wavelength is larger than 1 cm for ν_{res} = 24kHz and the mean free path is about 60 nm for p = 1000 mbar and about 60 μm for p = 1 mbar. It can be seen, the Q-value shows a quite different behaviour.

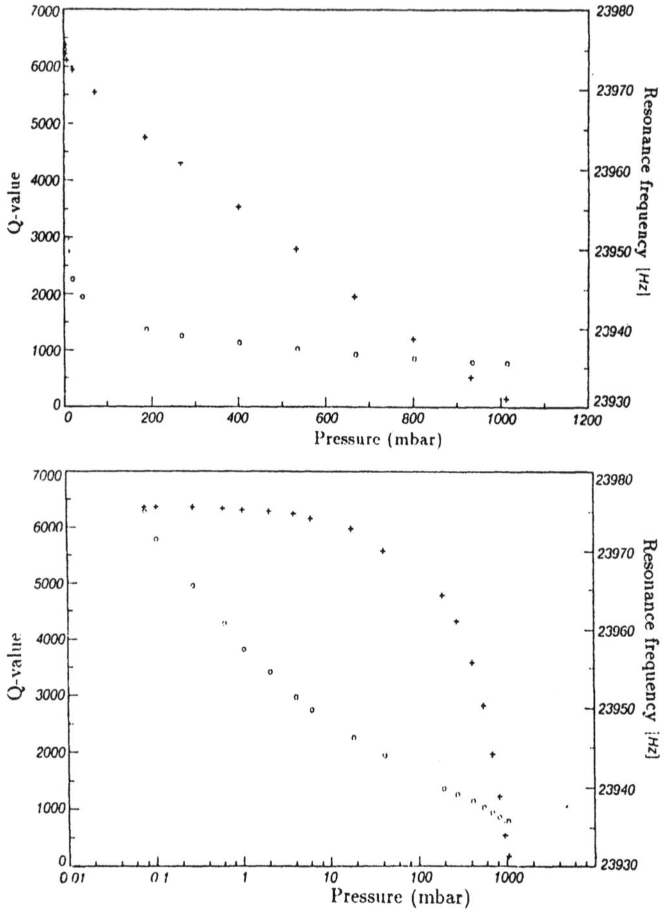

Fig. 6: Q-Value (o) and resonance frequency (+) versus pressure (linear and log. scale)

Temperature-Dependence

Fig. 7 shows the temperature-dependence of the resonance frequency. A linear decrease with T is obtained. The T-dependence is determined by

- the T-dependence of the elastic constant of Si
- the change of resonance frequency due to thermal expansion of Si
- the T-dependence of density and viscosity of the environmental medium (air usually)

In practice at room temperature its influence is negligble.

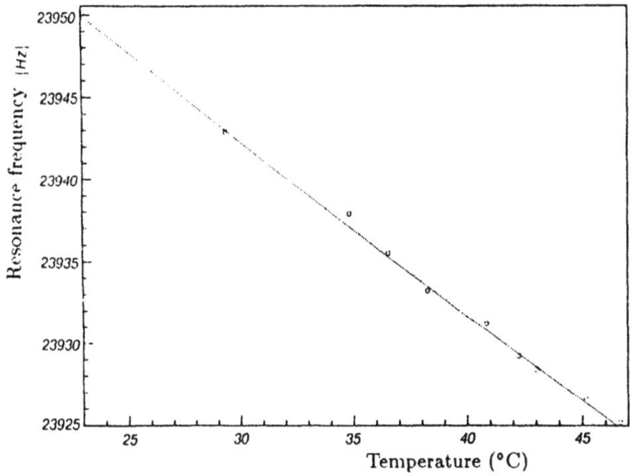

Fig. 7: Resonance frequency as function of temperature

Tip geometry

The tip geometry, however, is of tremendous influence, because not only the lateral resolution but also the tip-surface interaction is influenced by its radius of curvature and slope angle. But these values are not always very easy to determine. Therefore an ""in-situ"' measuring method is interesting. If a δ-function like tip is available, the tip under test can be characterised. Fig. 8 shows the result, a shows the tip under test and b shows a REM picture of the characterising tip. Very interesting are also well known test structures. Fig. 9 shows a periodic structure which has been performed by IBM with lithographic methods, as they are used in microelectronics.

Interaction between Tip and Sample Surface

The interactions between tip and surface can be artificially devided into short ($\leq$ 1nm) and long ($\geq$ 1nm) reaching ones. The shorter ones are better measured with the static the others with the dynamic method.

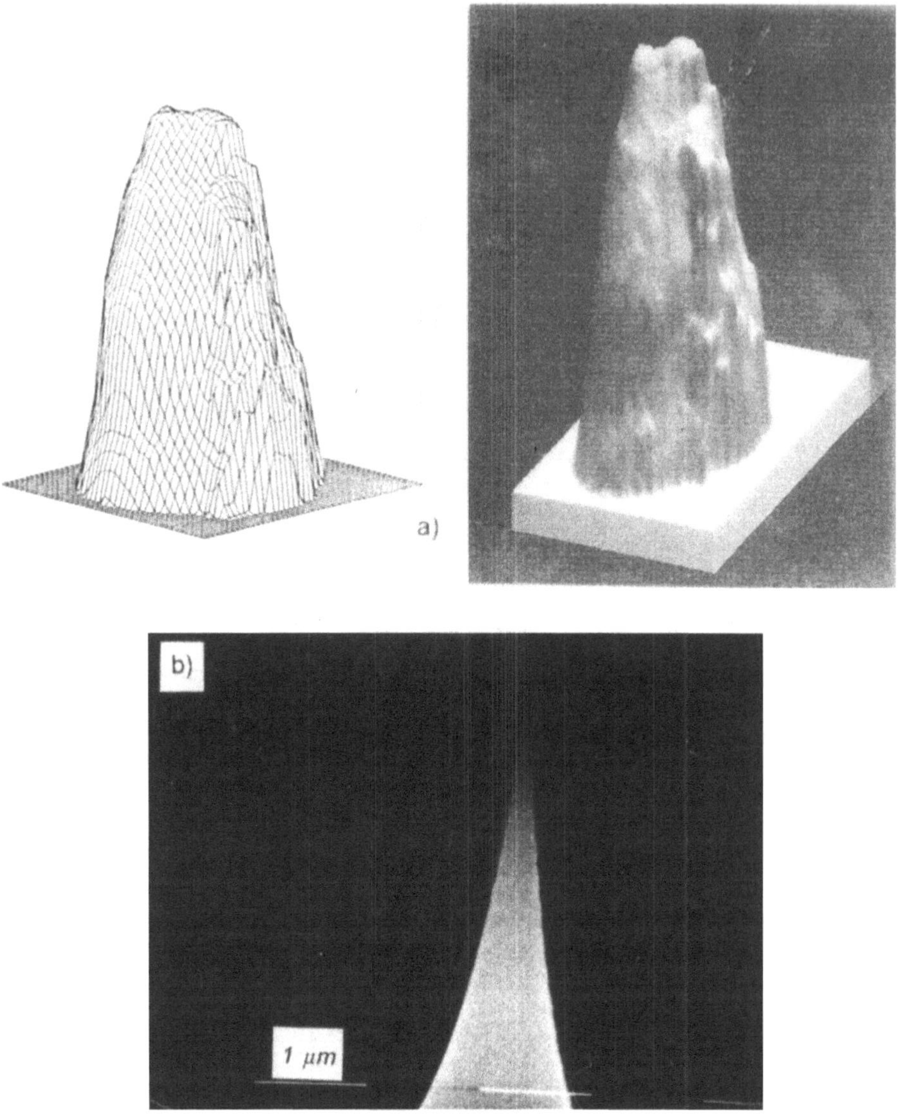

Fig. 8: SFM picture of a tip with a test tip a), area is 1 μm x 1 μm, tip height 2,5 μm, b) REM picture of the test tip showing the δ - function like shape

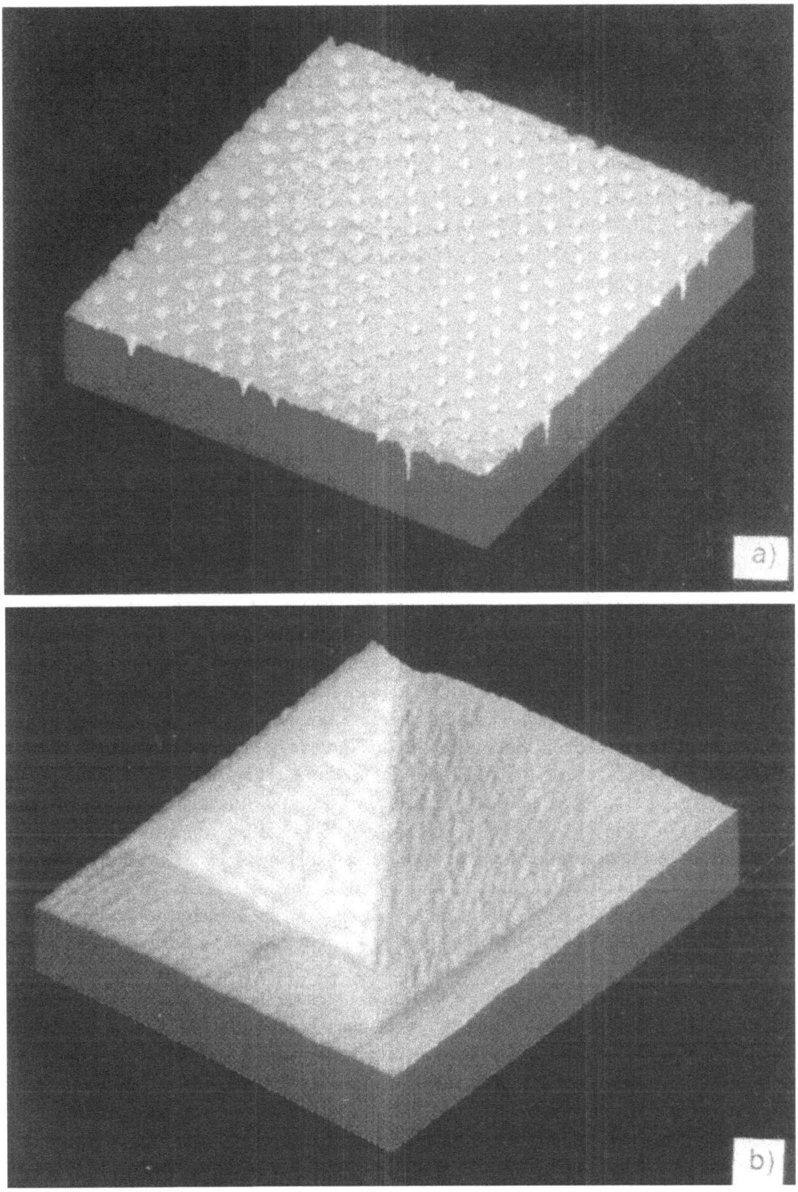

Fig. 9: SFM picture of a test structure a) general view (3 μm x 3 μm), b) detailed view (80 nm x 80 nm)

Hydrodynamic Interaction

As already mentioned, is the Q-value of an in air free oscillating cantilever totally determined by the damping of the air. Approaching the cantilever to a sample surface results in an additional damping due to the viscous friction of the between cantilever and surface displaced air. This damping is, of course, depending of the distance between tip and sample surface. Simultaneously with the Q-value the resonance frequency is changed.

Again assuming that the air can be treated as a viscous incompressible fluid the influence on the damping can be calculated if one further assumes a) that the velocity of the flowing air is zero at the cantilever and sample surface, and b) that the pressure is only x but not z-dependent. One obtains

$$Q(d) = \frac{Q_o}{1 + \frac{3\pi^2 \eta Q_o \nu_{res} L^4}{2Kd^3}} \tag{5}$$

with Q_o = Q-value of the undamped cantilever, η = friction coeffizient, L = length of cantilever, k = force constant, d = cantilever thickness. Fig. 10 shows the result of a measurement with a cantilever where instead of the tip a plate with effective dimensions 100 x 300 μm has been fixed. The cantilever data are 1100 x 100 x 5,5 μm, k = 0,5 N/m, $\nu_{res} \approx$ 6400kHz.

The solid line was calculated according to Equ. 5. It can be seen that the agreement is excellent. The Q-value already decreases at large distances in the range of the cantilever width of about 100 nm. This effect becomes the more apparent the lower the force constant of the cantilever is. At very low distances Q decreases to 1. The resonance frequency changes at much lower distances, a minimum is obtained at $\approx$ 20 nm and a strong increase at lower values of the distance. The reason for this increase is, that at these small distances an additional damping occurs due to the fact that the air does not totally flows out and in again. It can be estimated that at a distance of about 5 μm nearly 99,9 % of the air flows out and in but the rest of 0,1 % results already in a shift of the resonance-frequency of 600 Hz. The result of similar investigations but using instead of the plate a tip, that means the cantilever oscillates against a tip instead oscillating against the sample surface, are shown in the lower Fig. It turns out, that in air the influence of the tip can be neglected if the tip radius of curvature is smaller than $\approx$ 100 nm and the opening angle is small. Concluding one can say, the hydrodynamic interaction is strongly dependent on the cantilever geometry and the tip-surface distance. The smaller the cantilever dimensions the smaller the hydrodynamic influence. As a simple rule one can use for k values $\leq$ 1 N/m the tip height should be at least half of the cantilever width, for k $\geq$ 1 N/m correspondingly smaller. If deep structures are measured the hydrodynamic interaction can become of great importance because distance and effective area are changing. But in practice the hydrodynamic interaction can be of great help for the coarse approach of the tip to the sample surface without using a microscope, if the distance dependence is known.

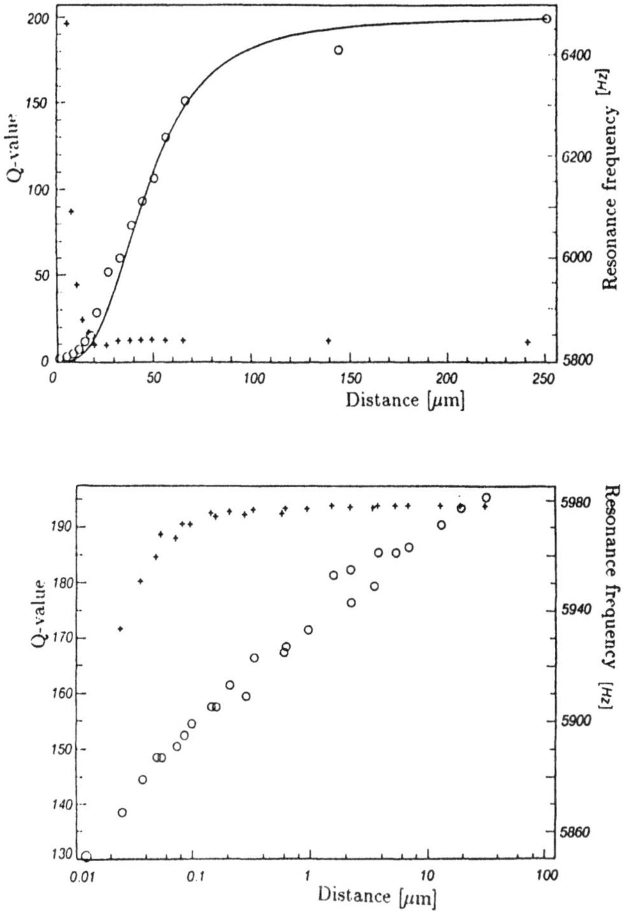

Fig. 10: Influence of the hydrodynamical interaction a) between cantilever and substrate surface on Q-value (o) and resonance frequency (+), solid line calculated, and b) between tip and sample surface

Electrostatic Interaction

The electrostatic interaction is due to the coulombic interaction between charges on tip and sample surface. Using conductive materials via the application of a tip bias this interaction can be changed, which offers new measuring possibilities. The electrostatic interaction can be described by the following relations

$$F_{el} = \frac{1}{2}\frac{dC}{dz}(U_k + U_p)^2$$

$$\Delta F_{el} = \frac{1}{2}\frac{d^2C}{dz^2}(U_k + U_p)^2 \qquad (6)$$

with U_k the contact potential, U_p the bias, z the tip to sample distance and C the effective capacitance.

Varying the bias and distance allows to determine the capacitance as well as the contact potential. The determination of the contact potential equals the well known Kelvin probe methode, but in the SFM methode the force or forcegradient are directly measured in form of a shift of the resonance frequency as a function of bias. This is shown in Fig. 11, the distance tip to sample surface has been 50 nm. A parabolic-function is obtained according to Equ. 6. From the shift of the zero-point of the curve the contact potential can be determined. An other method basing on the same relations is to use lower frequencies than the resonance frequency and to restore the resonance frequency by the applied bias. This is shown in Fig. 12 for experimental and calculated curves. If one measures the decrease of the amplitude of the oscillations varying the tip to sample distance with bias as parameter, one expects the steepest slope if F_{el} disappeares, i. e. the bias cancels the contact potential. Fig. 13 shows the result of such a measurement. It is seen that there is still a damping mechanism even if F_{el} disappears.

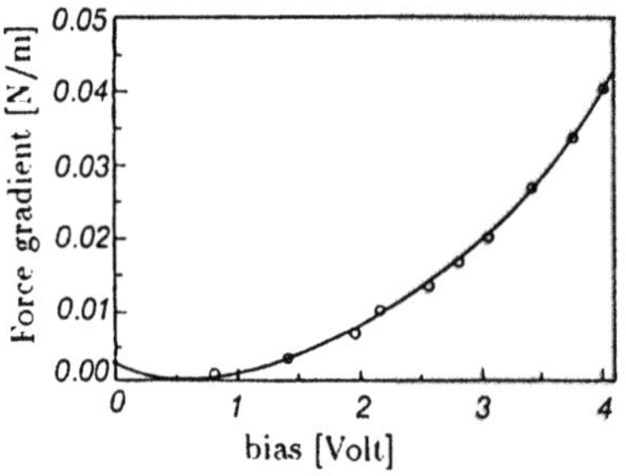

Fig. 11: Electrostatic force gradient versus sample bias

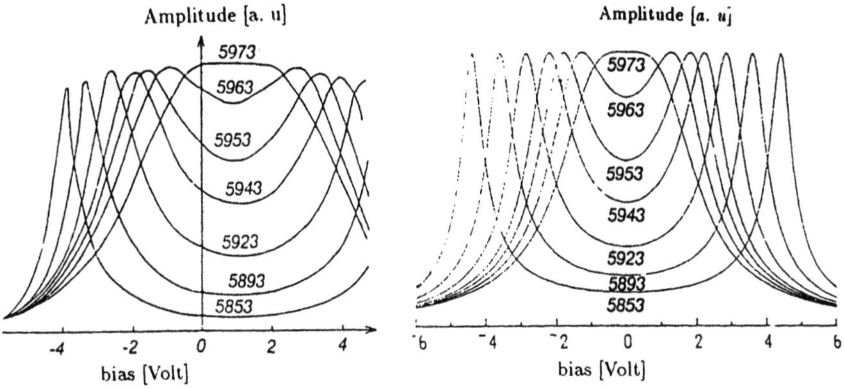

Fig. 12: Amplitude versus bias for different frequencies a) (resonance frequency 5975 Hz) and b) calculated curves of a)

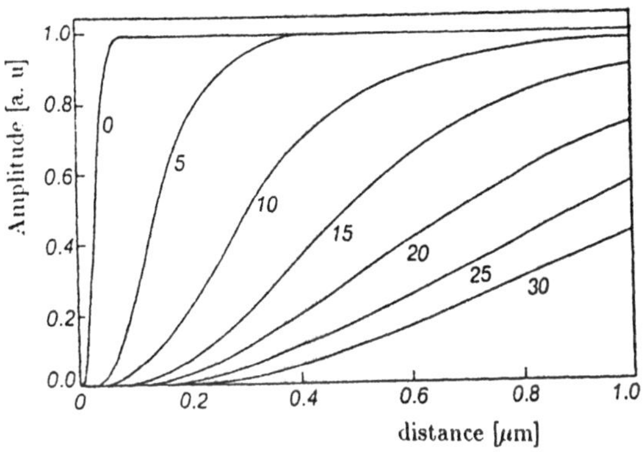

Fig. 13: Amplitude versus distance for different bias (0 - 30 Volt)

Van der Waal's Interaction

The van der Waal's interaction is caused by the interaction between dipoles. These dipoles may be already present in polar media or they are induced. They belong to the long range interactions which can be of importance up to 10 nm and more. They are complicated to calculate and non additive. In the simple case of induced dipoles without taking the retardation into account one obtains for the attractive force of two dipoles

$$w(r) = -\frac{3}{4}\frac{\alpha^2 I}{4\pi\epsilon_o^2}\frac{1}{r^6} = -\frac{C_{vdW}}{r^6} \tag{7}$$

the well known r^{-6} dependence, with I = the ionisation energy, α = polaribility, ϵ = permittivity.

To derive from this relation the interaction between tip and a planar sample surface by integrating over all the forces of the individual atoms of this plane, i. e. assuming additivity of the forces, which is not quite correct, one obtains

$$w(z) = \frac{\pi^2 C_{vdW} \rho_1 \rho_2 R}{6z} = -\frac{AR}{6z} \tag{8}$$

with ρ_i = density, and A = Hamaker constant. A more sophisticated discussion would be necessary, but is impossible in this paper.

In practice it is complicated to exclude the influence of the other interactions especially that of the electrostatic interaction. For this reason a bias has to be applied to compensate the contact potential. Fig. 14 shows an experimental result of Si-tip on Si surface, a) and b) are measured at the resonance frequency, c) and d) at insignificant reduced frequency, as a function of the tip to sample distance. The bias has been chosen to compensate the contact potential. If the distance is larger than about 2 nm, no hysteresis between approaching and retarding the tip occurs. If the instabel distance ≈ 2 nm is reached, however, that means where the attractive force is the same as that determined by the force constant of the tip, the tip stickes to the surface and a retarding force of about 400 nN was necessary to lift off the tip. The same happens if the frequency is lowered. In this case the resonance condition is reached at a certain distance.

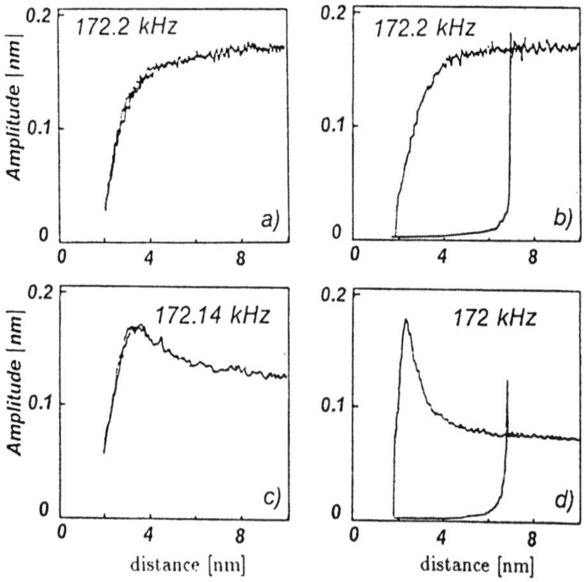

Fig. 14: Amplitude versus distance for different frequencies showing the influence of capillary forces. Si-tip on Si-surface.

The calulated curves are shown in this figure. The qualitative agreement is excellent but

- the unstabel point in the experiment is at about 2 nm whereas the calculation yields 0,5 nm
- the "'sticking force"' in experiment was about 400 nN where as the calculated one is about 20 nN
- the lift-off process in experiment is continously where the calculation yields an abrupt one.

The Capillary Interaction

Measuring in air or a gaseous medium it is possible that a bridge of fluid is built between tip and surface, which due to capillary forces changes the amplitude of the oscillations drastically. Fig. 15 shows a typical result and the model used for calculation.

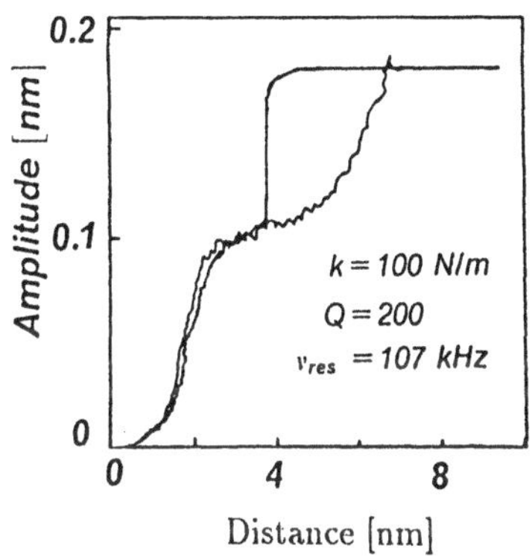

Fig. 15 a: Hysteresis occuring when approaching and retarding the tip below the critical distance, showing the influence of the capillary forces

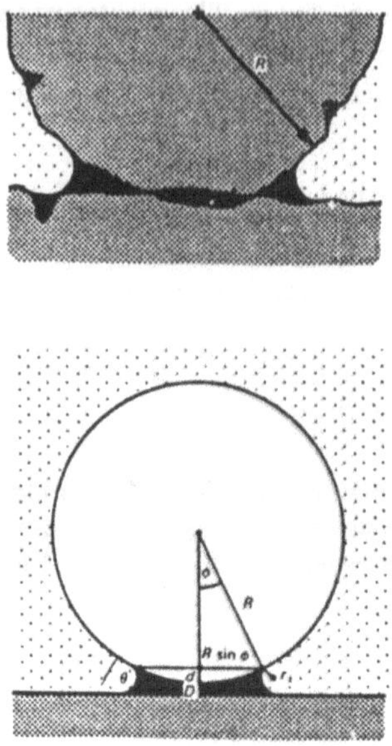

Fig. 15 b: Basic model for the calculation of the capillary forces

In a qualitative discussion, it is assumed that the tip is an incompressible sphere. If this tip approaches the surface, condensation takes place at a certain distance depending on the vapour pressure and concentration in the gas phase. In equilibrium the meniscus radius as a function of the relative vapour pressure p/p_s is given by the Kelvin equation

$$\left(\frac{1}{r_1}+\frac{1}{r_2}\right)^{-1} = r_k = \frac{\gamma M}{RT\log\frac{p}{p_s}} \tag{9}$$

with $\gamma$the surface tension, and M the molar volume, r_k the Kelvin radius. For water $\frac{\gamma M}{RT} = 0{,}54$nm. Therefore one obtains e. g. for $\frac{p}{p_s} = 0{,}9$ $r_k = 10$nm, for $\frac{p}{p_s} = 0{,}5$ $r_k = 1{,}6$nm and for $\frac{p}{p_s} = 0{,}1$ nm $r_k = 0{,}5$ nm.

The attractive force on the tip can than be estimated from the Laplace pressure:

$$P = \frac{\gamma}{r} \tag{10}$$

This pressure acts on the effective surface area $\pi(R\sin\emptyset)^2$. Fig. 16 shows the calculation of the force gradient as a function of the distance with the relative vapour pressure as parameter - for pure water and a tip radius of curvature of 20 nm. It can be seen the force gradient is nearly independent of the distance but strongly vapour pressure dependent. At high vapour pressure the gradient is very small, but the resonance frequency shift will be large, because the mass of water is much higher than that of air, thus the additional oscillating mass is large.

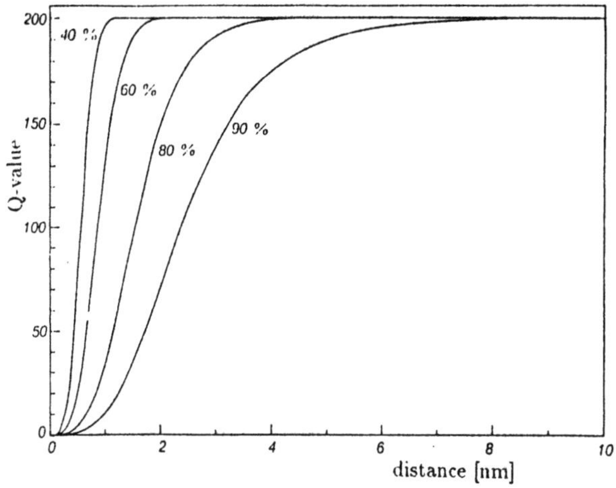

Fig. 16: Calculated influence of the capillary forces on the Q-value for different values of the relative humidity

The experimental result is shown in this figure. Pure water has been assumed hithertoo, but due to surface contamination this is not very realistic. Therefore the discussion can be qualitatively only.

In Fig. 15 a) at about 4 nm the amplitude changes very abrupt and remains than constant for some distance, to be reduced to zero due to the van der Waals interaction at lower distances. By retarding the tip a hysteresis occurs.

If the waterbridge occurs, the Q-value is drastically reduced. This can be qualitatively calculated assuming a similar behaviour of the fluid like the air but taking the larger mass into account.

This means for the controll system, that not anymore the shift of the resonance frequency but the damping is dominant for the distance dependence. This can be used for a stabel control behaviour, it means that during the scanning of the tip the waterbridge will be carried along. The tip radius of curvature is of great importance. This effect has to be taken into account also in the STM.

Influence of the interaction on the electronic control system

From the discussion of the 4 types of interaction it follows, that the electronic control system has to be adjusted. The van der Waals and electrostatic interaction are influencing in first approximation the resonance frequency only, whereas the hydrodynamic and capillary forces cause strong damping, i. e. reduction of the Q-value.

Using the long range hydrodynamic force and damping in air, only a small lateral resolution is obtainable ($\approx 1\ \mu m$) thus the application as an acoustic nearfield microscope is only possible with great restriction, or one uses media with very high damping effect.

Because in general the simultaneous action of all types of interactions have to be expected, an exact knowledge of the control-mechanism and the measuring parameters are necessary for a correct interpretation of the experimental results.

Characterisation of sample surfaces

Especially the electrostatic interaction offers new interesting possibilites. Because it is possible to change the tip to sample surface interaction via an applied bias it is possible to retard the tip by the bias, s. Fig. 17. This means topographic structures can be measured for different distances of tip to surface and these distances can be electrically controlled. The lateral resolution, of course, is reduced. But the tip can be retarded to a distance which is larger than the largest expected topographic structure of the sample, thus without any contact of the tip with the surface a very high scanning speed can be obtained of more than 100 $\mu m/s$ for a general mapping of the surface. Also capacitance and surface potential can be measured, applying a bias of $U = U_0 \sin \omega t$.

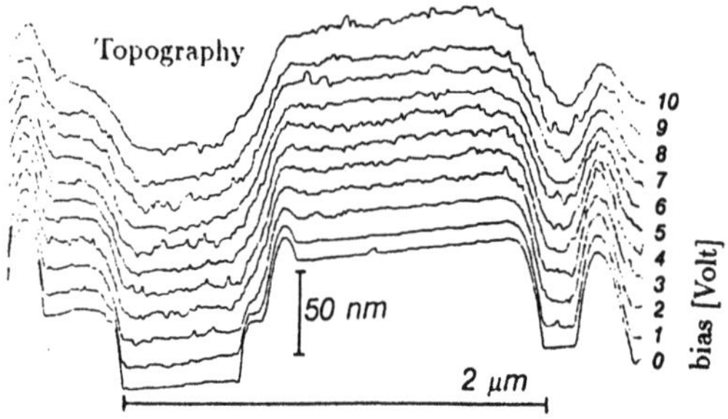

Fig. 17: SFM measurement of the topography for different bias values, suggesting the possibility of a high speed general mapping

Fig. 18: Simultaneous imaging of the topography (2,5 μm x 2,5 μm) (left), and the capacitance of a 1 μm SiO_2 structure on silicon (right)

Fig. 18 shows a 1 μm SiO_2 structure a) the topography and b) the capacitance, it is seen the structure is inverted. Fig. 19 shows the same picture, but single lines, and that lateral structures lower than 50 nm have been resolved. Fig. 20 shows the structure of Si_3N_4 on Si, topography and capacitance.

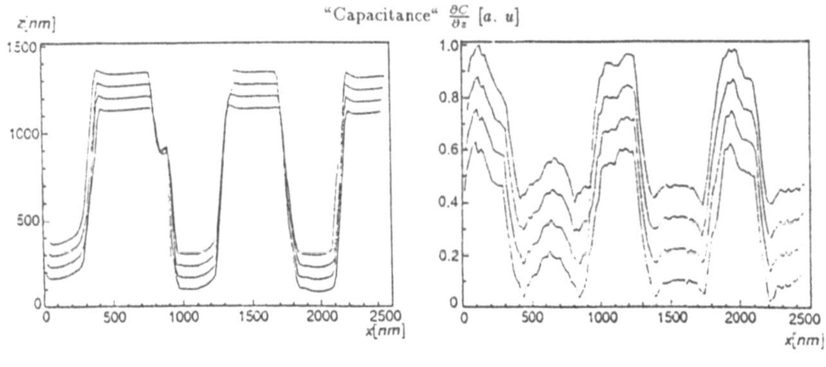

Fig. 19: Scanning of a deep trench (1 μm) in the topography and capacitance mode

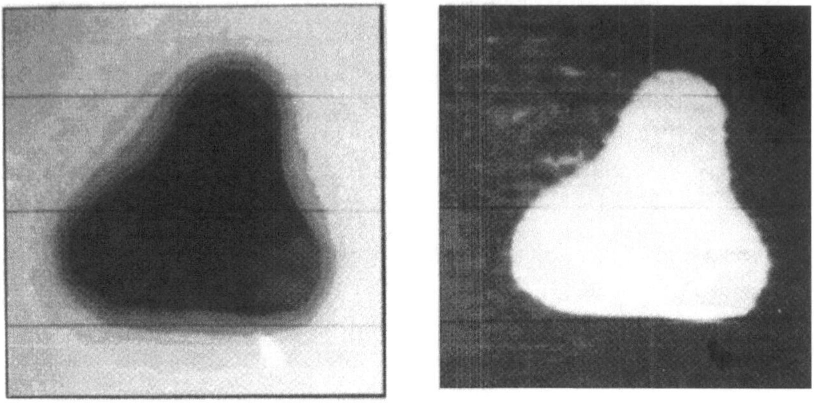

Fig. 20: Simultaneous measurement of the topography (left) and thickness (right) of a Si_3N_4 - film (25 nm) on Silicon in the electrostatic SFM mode

Acknowledgement:

This paper is part of the PhD Thesis of Dr. M. Nonnenmacher, which has been written at the Institute of Technical Physics of the University of Kassel and which has been greatly supported by IBM, Sindelfingen (Dr. Greschner, Dr. Wolter).

References

[1] [1] H. K. Wickramasinghe, "'Scanned-Probe Microscopy"', Scientific American, 98 - 105, Okt. 1989

[2] [2] G. Binnig, H. Rohrer, "'Scanning Tunneling Micr."', Helv. Phys. Acta 55, 726 (1982)

[3] P. Muralt and D. W. Pohl, Appl. Phys. Lett. 48, 514 (1986)

[4] P. Gthner, U. Ch. Fischer and K. Dransfeld, Appl. Phys. B48, 82 - 92 (1989)

[5] Y. Martin and H. K. Wickramasinghe, Appl. Phys. Lett. 50 (20), 1455 (1987)

[6] J. R. Matey and J. Blanc, J. Appl. Phys. 57, 1437 (1985)

[7] P. K. Hansma, B. Drake, O. Marti and C. B. Prater, Science 243, 641 (1989)

[8] C. C. Williams and H. K. Wickramasinghe, Appl. Phys. Lett. 49, 1587 (1986)

[9] G. Binnig, C. F. Quate, Ch. Gerber, Phys. Rev. Lett. 56, 930 (1986)

[10] M. Nonnenmacher, J. Greschner, O. Wolter, R. Kassing, Optik, Sup. 4 (Vol 83), 70 (1987)

[11] M. Nonnenmacher, J. W. Bartha, O. Wolter, D. Pohl, R. Kassing, "'Topographier- und Kraft-Messungen an dnnen Schichten - untersucht mit einem Raster- Tunnel- und Kraft-Mikroskop"' BEDO-Band 21, 13 - 18 (1988)

[12] Nonnenmacher, M.: PhD Thesis, University of Kassel 1990

New Scanning Microscopy Techniques: Scanning Noise Microscopy Scanning Tunneling Microscopy Assisted by Surface Plasmons

R. Möller
Fakultät für Physik, Universität Konstanz, 7750 Konstanz
Federal Republic of Germany

Abstract

The present paper will be divided into two sections: In the first part the principle of the scanning noise microscopy i.e. the combination with potentiometry will be discussed. The second part will describe the investigation of the interaction of surface plasmons with the tunneling process of electrons.

The analysis of the current noise in scanning tunneling microscopy experiments reveals that at zero bias white thermal noise is observed. Its amplitude is given by the Nyquist formula for the tunneling resistance. Since the latter depends exponentially from the gap distance it is possible to operate a scanning tunneling microscope at exactly zero bias between sample and tunneling tip. The electronic feedback for the gap distance maintains a constant current noise instead of operating at constant direct current. Tests on a polycrystalline silver surface show that the obtained resolution is comparable to the one in conventional STM. As already a minor deviation from zero bias between tunneling tip and sample surface leads to a significant dc-component of the average current scanning noise microscopy can be very well combined with potentiometry. For any point on the sample the potential can be accurately measured by adjusting the potential such that the dc-component of the average tunneling current vanishes. The resolution obtained in our experiment is in the order of 1 μV.

The influence of surface plasmons excited in a polycrystalline silver film on the tunneling current of a scanning tunneling microscope has been analyzed. The plasmons cause an additional flow of electrons from the tungsten tip to the silver surface on the order of up to 50 pA. This process is independent of the polarity of the applied bias voltage, thereby excluding effects of thermal expansion. The different nature of the ordinary tunneling current and the surface plasmon induced current is clearly revealed by their different dependence on the gap distance. The local distribution of the intensity of the surface plasmon induced signal reveals structures on a nanometer scale.

Scanning Noise Microscopy

The noise of the tunneling current in scanning tunneling microscope (STM) is limiting the resolution of the instrument. Different methods have been proposed to reduce its effect either by data processing [1] or by experimental techniques [2]. The subject of this article is to discuss the use of the current noise for new experimental methods.

The experimental setup has been described in detail in ref. [3]. The STM uses a spring reduction system for the coarse approach between tunneling tip and sample. The fine approach and the lateral scanning are realized by a single tube scanner. Vibrational damping is provided by a stack of metal plates separated by pieces of viton.

The tunneling tips have been prepared by electrochemical etching of a tungsten wire of 0.5 mm diameter. The tips have been cleaned by field desorption prior to the measurement. To prepare the sample about 40 nm of silver have been deposited on a stainless steel surface by evaporation. The preparation of the sample, the cleaning of the tip, and the measurement have been performed under UHV-conditions.

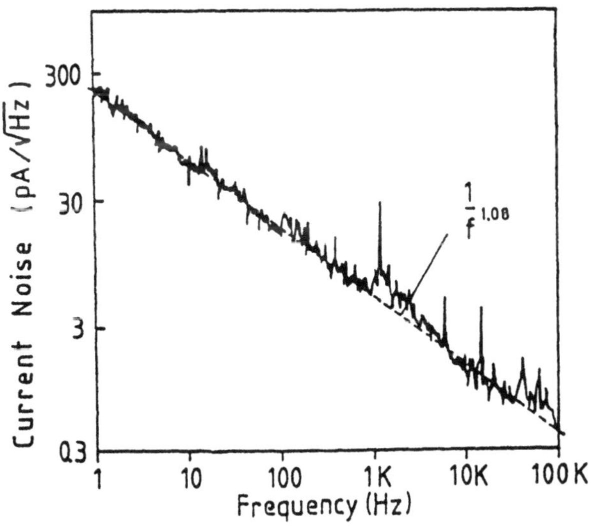

Fig.1 Spectrum of the current noise measured at fixed gap distance. The average current is 1 nA at a sample bias V = -0.25 V.

Fig.1 shows the noise spectrum of the tunneling current at an average current of 1 nA and a sample voltage of -0.25 V. The noise reveals a clear 1/f characteristic. It exceeds by far the shot noise limit given by 18 fA/√Hz for a tunneling current of 1 nA. The dashed line in the Fig.1 represents a fit to the data given by $1/f^{1.08}$. Although the current noise shows some variations on the experimental conditions the data plotted in Fig.1 are representative for typical conditions in STM experiments.

However, the noise spectrum is changed if the applied bias voltage is strongly reduced. Fig.2 a,b show the current noise for 0.5 mV and zero bias between tip and sample. Both spectra have been obtained at a gap distance corresponding to

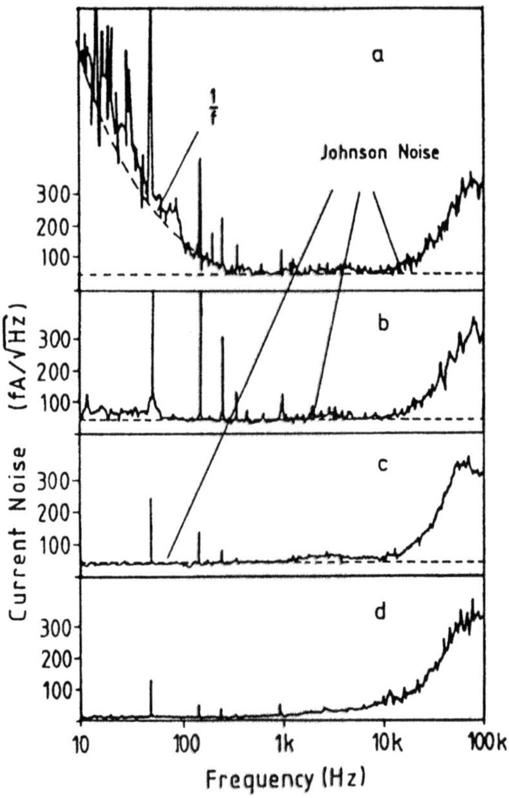

Fig.2 (a) Spectrum of the current noise at a tunneling current of $10^7\ \Omega$ and a sample bias of -0.5 mV. (b) Spectrum of the current noise at a tunneling resistance of $10^7\ \Omega$ and zero bias. (c) Spectrum of the current noise of a metal-film resistor of $10^7\ \Omega$. (d) Noise spectrum of the I/V converter with a gain of 10^9 V/A.

a tunneling resistance of $10^7\ \Omega$. This has been realized experimentally by tunneling at 10 mV bias and 1 nA current before the feedback loop for the gap distance has been blocked.

At a bias of 0.5 mV the noise spectrum has a 1/f contribution for the frequency range between 10 and 500 Hz. At zero bias the 1/f noise vanishes and white noise is dominant. This contribution can be clearly identified as thermal resistor noise given by the Nyquist formula, which leads to a noise amplitude of 41 $fA/\sqrt{Hz}$ for a resistance of $10^7\ \Omega$. For comparison the noise spectrum of a $10^7\ \Omega$ metal film resistor is displayed in Fig.2 c. Neglecting the perturbation at 50, 150, 250 Hz a good agreement between the data of Fig.2 b and Fig.2 c can be seen. To identify the contribution due to the measuring system the spectrum of the preamplifier for the current without any input is shown in Fig.2 d.

The analysis of the thermal noise at zero bias has been performed for various gap distances and it could be verified that it scales like $R^{-1/2}$, where R is the tunneling resistance [4]. Since the latter depends exponentially on the gap distance the amplitude of the thermal noise shows also an exponential distance dependence.

The observation that due to the thermal noise of the tunneling gap the distance between the tunneling tip and the sample can be monitored even at zero bias suggests a new type of scanning microscopy. If the z-position of the tip is adjusted such that the amplitude of the current noise remains constant the gap distance can be controlled. The method is analogous to conventional STM, however the tunneling current is replaced by the noise amplitude.

An interesting application of the "scanning noise microscopy" comes about if one considers the information contained in the tunneling current at almost zero bias. While the current noise and thereby the gap distance manifests in the ac-component, a difference in the electric potential between the tip and the sample leads to a dc-component of the current. By means of a second feedback loop acting on the tip potential the potential difference can be eliminated. With this technique the topography and the local electric potential of a sample surface can be measured simultaneously.

Fig.3 shows the scheme of the electrical circuitry. The thermal noise and an eventual dc-component of the tunneling current at the sample are detected by a very sensitive low-noise current-to-voltage converter. Its output is connected to two separate feedback loops, one for the adjustment of the distance between tip and sample and the other for the correction of the electric potential between tip and sample.

To control the distance between tip and sample the ac-component of the detected current representing the thermal noise is filtered by a lowpass filter (cut-off frequency 10 kHz) to suppress electronic noise of the converter. It is rectified and by means of a logarithmic amplifier converted to a signal which is linearly proportional to the gap distance. The difference of this signal to a preset value is integrated and fed into the high-voltage amplifier for the z-electrode of the piezo. As in scanning tunneling microscopy this signal represents the z-value for the topography.

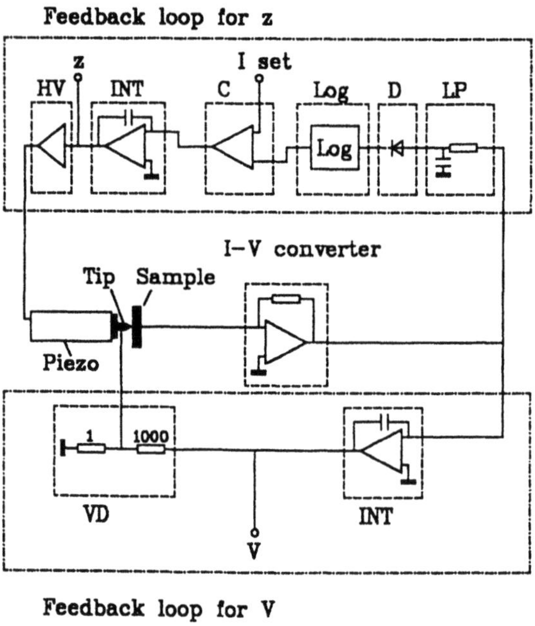

Fig.3 Scheme of the electric circuitry for the "scanning noise potentiometry". The following abbreviations are used:
HV high voltage amplifier, INT integrator, C differential amplifier, Log logarithmic amplifier, D rectifier, LP low pass filter, VD voltage divider

The second feedback consists of an integrator and a voltage divider. If a dc-component were present in the tunneling current the potential of the tunneling tip is corrected by the integrator until the average current vanishes. The voltage divider 1:1000 adapts the useful range of the integrator to the investigated range for the potential differences which is in the range of a few hundred μV. It also helps to reduce unwanted electronic noise. The output of the integrator provides the local electric potential (scaled by a factor of 1000) and is used for the data acquisition. The signals for the topography and for the local electric potential are recorded simultaneously by a computer.

Fig.4 and Fig.5 show the topography and the local electric potential for the polycrystalline silver surface; no voltage is applied across the sample. The area is 500 x 500 nm, the maximum difference in height is 15 nm, the maximum difference in the potential amounts 56 μV. The average current noise amplitude is about 1 pA. It can be seen that the different facets of the silver crystallites are well resolved.

In the representation of the local potential in Fig. 5 the bright areas correspond to a negative potential and vice versa. By comparing subsequent lines the resolution is estimated to be about 1 μV. Qualitatively the potential appears like the high pass filtered signal of the topography. The critical length for the cut-off is in the order of 50 nm.

Different tests have been performed to check the reproducibility of the measurement of the local electric potential. The result is independent of the direction of the scan or the average amplitude of the current noise for the range of 0.5 to 5 pA. Furthermore, the influence of a small offset current on the results has been checked by introducing a voluntary offset in the order of a few 100 fA. However this did not lead to a significant change of the results.

A tentative explanation of the observed variation of the local electric potential can be given by regarding the grainy structure of the silver film. In Fig. 4 it can be seen that the surface consists of microcrystallites showing atomically flat facets. Since the work function Φ of the silver surface depends on the crystallographic orientation the observed electric potential which is given by the difference between the work function of the tip and the sample should vary between the different facets. For the (111) and the (100) face Φ-values of -4.75 V and -4.81 V have been reported [5], i.e. a difference of 0.06 V. For the high-index faces with less dense packing lower values are to be expected. The observed variation of the electric potential of up to 50 μV is in the right order of magnitude considering that the charge distribution is smeared out over small dimensions.

Fig. 4 Topography of a polycrystalline silver surface measured by "scanning noise microscopy" at zero bias between tunneling tip and sample surface. The area is about 500 x 500 nm. The total difference in height is about 15 nm.

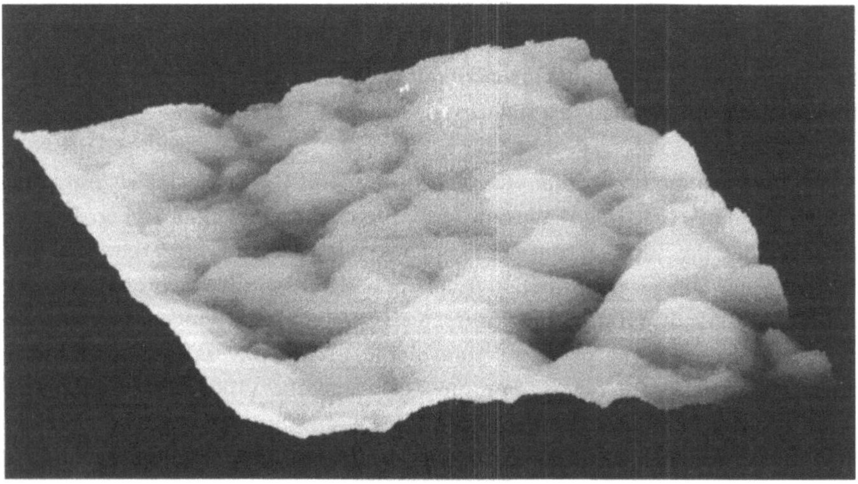

Fig.5 Local electric potential corresponding to Fig.4 measured by "scanning noise potentiometry". The maximum potential difference is about 56 μV.

The noise of the tunneling current of an STM has been analyzed. At zero bias the noise is given by thermal noise of the tunneling resistance. This can be used to operate a scanning microscope without applying a bias voltage between tip and sample. The distance can be controlled by regulating on a constant noise amplitude. This "scanning noise microscopy" can be very well combined with potentiometry [6] leading to a very good resolution as well of the topography as of the local electric potential. The resolution of the electric potential of about 1 μV is comparable to most advanced techniques of potentiometry [7]

The method requires only a simple electronic setup and provides a scan speed which is only slightly lower than the ordinary STM operation. Depending on the purpose of the experiment the presented technique can be used to analyze different features. If a voltage is applied across the sample the current flow and the associated voltage drop can be determined. In case of nonuniform heating it is possible to evaluate the temperature distribution on the sample with high accuracy [8] (e.g 0.02° K for tungsten and gold as tip and surface material). If the surface consists of different chemical species their arrangement on the surface can be analyzed due to different thermovoltages by heating the sample[9].

It is demonstrated that this technique permits to resolve a local variation of the contact potential on a polycrystalline silver film evaporated on a stainless steel substrate. Potential differences of up to 50 μV have been observed. The observation is attributed to changes of the work function associated with the different crystallographic orientation of the facets of the silver grain.

Scanning Tunneling Microscopy Assisted by Surface Plasmons

The interaction of visible light with electrons tunneling through a vacuum gap is of fundamental interest. Different mechanisms for this interaction have been proposed and some of them have been verified experimentally. Amer et al. [10] reported the thermal expansion of the tunneling tip and possibly the sample surface upon radiation with visible laser light. The use of the nonlinearity of the current voltage characteristic for the direct rectification of light has been discussed by Cutler et al. [11]. They proposed to determine the traversal time for tunneling of the electrons by the cut-off frequency above which the rectification process would no longer operate. Mixing of two laser frequencies in the tunneling gap has been demonstrated by Krieger et al. [12] and Arnold et al.[13] for frequency differences ranging from MHz up to several THz. These results have been explained by thermal expansion of the outermost part of the tip at low frequencies and by the nonlinearity of the current-voltage characteristic at high frequencies. On semiconducting surfaces a photovoltage in the order of a few 100 mV has been observed upon radiation with laser light [14].

Gimzewski et al. [15] have observed the emission of light from a tunneling junction when tunneling on silver surfaces. This effect has been attributed to the excitation of surface plasmons by inelastic tunneling of electrons [16]. Here the observed photons result from the decay of the plasmons. Recently the detection of luminescence from III-V heterostructures during the tunneling process with nanometric resolution has been reported [17].

The experiment presented in this paper involves the inverse process. Surface plasmons optically excited on a silver film are shown to induce an additional current of electrons from the tip to the sample. This current is superimposed on the tunneling current observed in absence of surface plasmons.

The experimental setup allows for the first time to investigate the spatial distribution of the AC-electric field induced by the plasmons with a resolution in the range of nanometers. This may help towards a better understanding of the processes leading to surface enhanced spectroscopy effects, where local field enhancement at specific sites at the surface (e.g. at points with large curvature, pores below the surface or at sites of chemisorption) is believed to play an important role [18].

The experimental setup is sketched in Fig.6. The STM used for the experiment has been described in detail elsewhere [19]. The coarse approach is performed by piezo driven micrometer screws allowing a precise positioning of the tunneling tip relative to the sample in all three dimensions. Vibrational insulation is provided by a stack of metal plates separated by pieces of viton. The tunneling tips have been prepared by electrochemical etching of a tungsten wire of 0.5 mm diameter with 2 mol l-1 KOH.

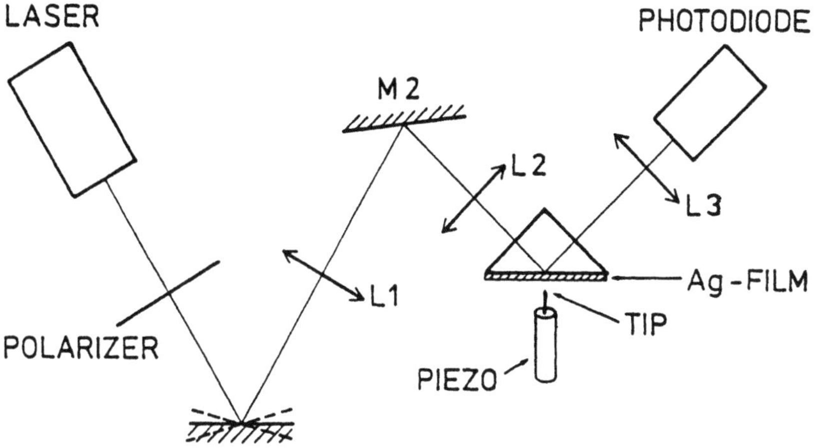

Fig.6 Schematic representation of the experimental setup

The surface plasmons are optically excited in a Kretschmann configuration [20]. The light of a 5mW HeNe-laser is scanned at a frequency of 1.3 kHz over a small angle by a mirror M1 (see Fig.6) mounted on a bimorph piezo. A telescope optical arrangement (lenses L1 and L2) then images the angularly modulated beam onto a spot of about 0.5 mm diameter on the silver film, thermally evaporated onto the prism base. The mirror M2 allows for positioning of the spot. The attenuated total reflection signal is collected by a lens (L3) and detected with a photodiode.

Fig.7 shows synchronized traces of the signal on the photodiode (a) and the tunneling current for a tip voltage of +50 mV (b) and -50 mV(c). The excitation of surface plasmons is characterized by the "dips" in the signal of the photodiode due to the attenuated total reflection. It can be clearly seen that a correlated effect is observed in the tunneling current. The absolute value of the average tunneling current has been set to 20 pA for both curves (b) and (c). The positive sign of the current corresponds to electrons tunneling from the sample to the tip. For positive bias at the tip the tunneling current is reduced when the surface plasmons are excited.

For the particular parameters of trace (b) the current is even reversed. At negative tip bias the tunneling current is enhanced. The relative intensity of the plasmon induced signal (in the following named PS) is smaller, since for the given I-V characteristic (Fig. 10b) the gap distance is increased at negative tip bias if the absolute value of the tunneling current is maintained.
On the observed time scale (the plasmon excitation lasts about 70 μsec) the effect appears to be instantaneous.

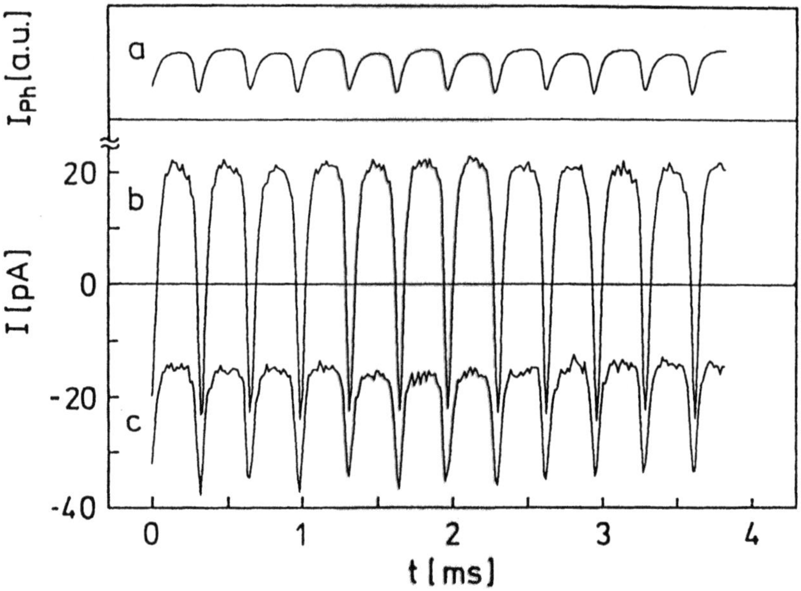

Fig.7 Correlation between the excitation of surface plasmons and the tunneling current
(a) signal of the photodiode showing the attenuation of the total reflection if surface plasmons are excited, (b) tunneling current at a tip bias of + 50 mV, the average value of the tunneling current is 20 pA, (c) tunneling current at a tip bias of - 50 mV, the average value of the tunneling current is - 20 pA

Since the sign of the PS is independent of the bias voltage between tip and sample it can be excluded that the observed effect is caused by thermal expansion. The latter always leads to an increase of the absolute value of the tunneling current.

To get more insight into the nature of the coupling between the surface plasmons and the tunneling current the latter has been measured as function of the gap distance between tunneling tip and sample surface. Fig.8 shows the PS (a) and the tunneling current (b) as function of the gap distance. While the tunneling current decreases exponentially the PS signal shows a complex behavior. It reveals only a weak distance dependence in the range of typical tunneling parameters. If the gap distance is further increased the decrease of the PS becomes stronger. However, even at the maximum gap distance in Fig. 8 where no tunneling current was detectable in the absence of surface plasmons the PS still amounts to about a third of the value at the typical operating conditions.

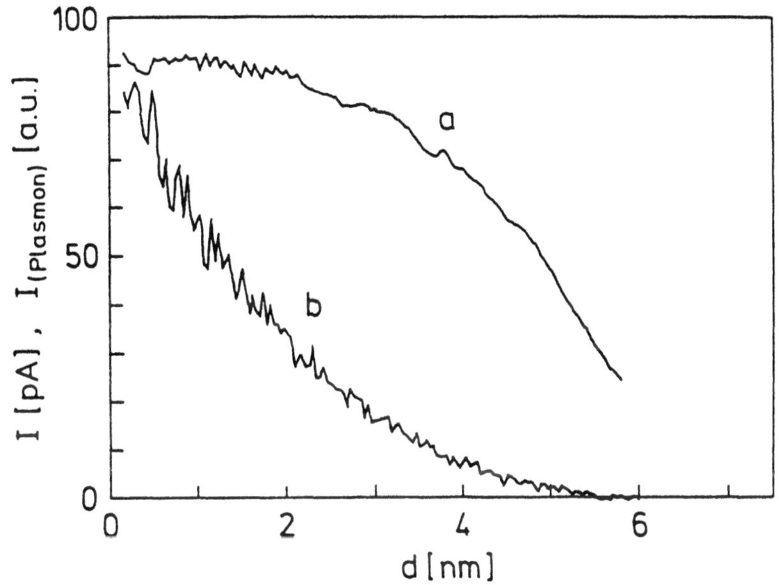

Fig.8 Plasmon induced signal (PS) and tunneling current as function of the distance between tunneling tip and surface sample. (a) plasmon induced signal (PS), (b) tunneling current in absence of plasmons

This distance dependence leads to the conclusion that the observed effect is due neither to a surface photovoltage nor to a thermovoltage caused by differential thermal heating. In both cases a simple exponential dependence is to be expected as the photovoltage or the thermovoltage act as constant voltage source in series with the applied bias.

The data displayed in Fig.7 and Fig.8 have been measured at a particular site on the sample showing a strong PS. To investigate its local intensity distribution the PS has been recorded simultaneously with the topography of the silver surface observed by STM. The upper part of Fig. 9 shows the topography of an area of about 60 x 60 nm obtained at a tip bias of 150 mV and a tunneling current of 20 pA. To measure the PS the output of the current-to-voltage converter amplifying the tunneling current is connected to a lock-in-amplifier. The angle of incidence of the laser beam is varied sinusoidally. Since the plasmons are excited at the center of the scan range, the plasmon signal can be detected by using the second harmonic of the scan frequency as reference for the lock-in-amplifier. The correlated PS is displayed in the lower part of Fig. 9. It can be seen that for a large part of the surface the signal is weak and only on particular areas a strong effect appears. The weak signals correlate with the topography but no particularity of the sites exhibiting a strong PS can be recognized in the topography.

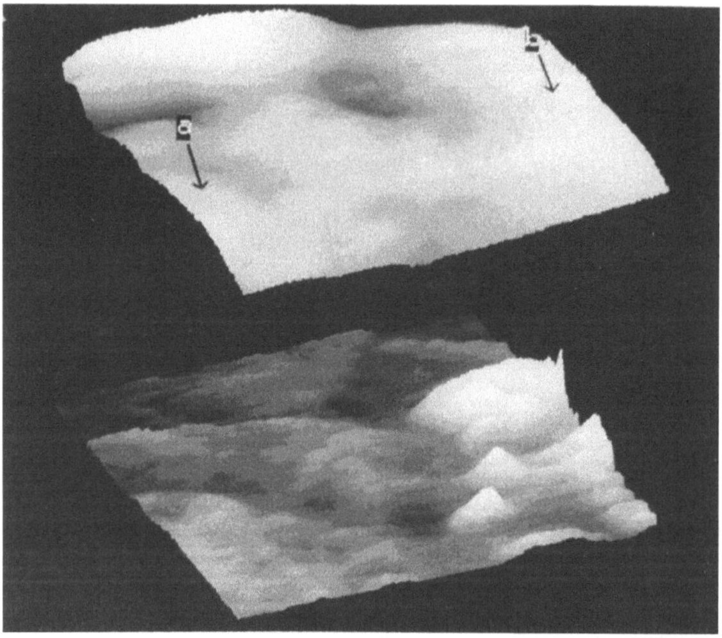

Fig.9 Simultaneous recording of the surface topography and the plasmon induced signal (PS). The investigated area is about 60x60nm, the tip bias is 150 mV, the tunneling current is 20 pA. Two positions at which the I-V characteristics displayed in Fig.10 have been measured are marked (a) and (b). The upper part shows the topography observed by STM; the lower part gives the corresponding plasmon induced signal.

To understand the local variation of the PS the tunneling current has been measured as function of the applied bias voltage for a site showing no PS (marked a in Fig 9) and one with a strong PS (marked b). The resulting I-V characteristics for V varying from -150 mV to +150 mV are displayed in Fig. 10. At the position marked (a) a nearly linear dependence is found. A very different curve is found at (b) where strong PS is observed. The characteristic shows a strong nonlinearity; a strong increase of the absolute value of the tunneling current is found at negative tip bias. The surface plasmons are excited periodically, the dips in the curve represent the PS.

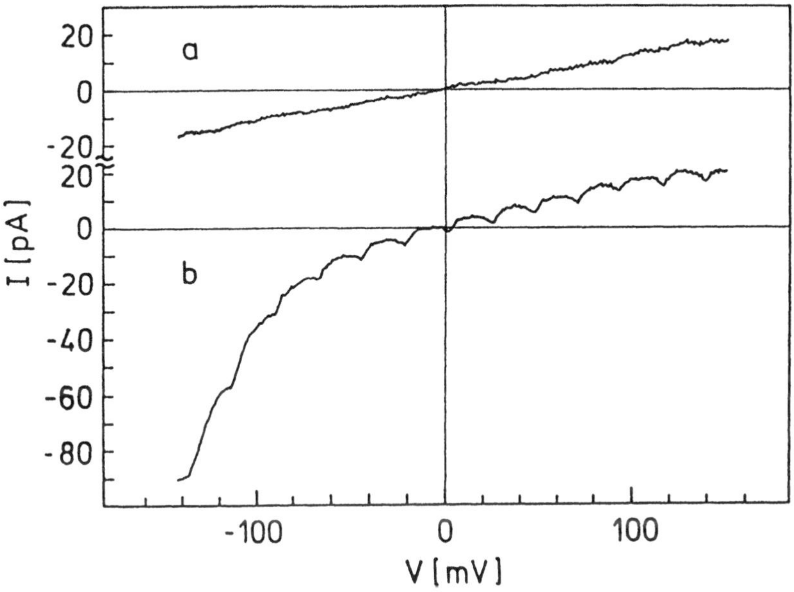

Fig.10 Comparison between two different current-to- voltage curves
(a) linear I-V curve measured at the position (a) in Fig.4 where no strong PS has been detected, (b) nonlinear I-V curve measured at the position (b) in Fig. 4 where a large PS has been detected

The nonlinear form of the I-V curve appears to be characteristic for the spots showing a strong PS. This behavior has been found whenever PS is observed. The nonlinearity of the I-V curve represents a necessary condition, a PS is, however, not always observed if a nonlinear I-V curve is measured.

The experimental findings strongly suggest that the observed effect is due to the rectification of the AC-field of the surface plasmon by the nonlinearity of the I-

V dependence of the tunneling gap. Although it is questionable whether the quasi-static I-V characteristic measured at low frequencies is applicable as well for optical frequencies of $5x10^{14}$ Hz, the experimental results agree well with the conclusion drawn from the static I-V curve. Therefore the PS should increase with the "curvature" of the I-V curve i.e. if a strongly curved part of the I-V curve is selected by choosing the corresponding bias. Measurements performed with the lock-in-amplifier reveal that the PS increases by 50 % if the bias at the tunneling tip is changed from +50 mV to -50 mV at constant gap distance.

We assume that the distance dependence of the PS is to a great part due to changes of the I-V characteristic as a function of distance. This might also explain the complex distance dependence discussed above. For small gap distances the relative increase of the curvature with growing gap distance then compensates for the overall decrease of the tunneling current.

The excitation of surface plasmons in a silver film leads to an additional flow of electrons from the tunneling tip to the silver surface. Since the PS is only observed if the I-V characteristic shows a strong nonlinearity it is concluded that the observed signals are due to rectification of the AC-field of the surface plasmon. Using the static I-V curve one can estimate the AC-voltage across the gap which is required to generate the observed signals. For a strong PS values of about 100 mV are obtained. The effect of rectification of an AC-field by the nonlinear I-V curve of the tunnel junction has been discussed by Cutler et al. [11] for the case that the laser light is shining directly onto the junction. The distance dependence of the PS found in our experiment shows a good agreement with their theoretical prediction.

However, our results do not confirm the theoretical prediction by Tagliacozzo et Tosatti [21] that surface plasmons should lead to a lowering of the potential barrier for the tunneling electrons resulting in an increase of the tunneling current independent of the polarity of the applied bias voltage. If such effect is present it has to be weak in comparison to the rectification discussed above.

The local distribution of the PS varies dramatically. It is assumed that the large signals are observed if the I-V characteristic is strongly curved and if a strong local electric field is present. The experiments have been performed in air, hence the I-V curve can be completely changed by impurities. However, it should be possible to prepare the sample and the tunneling tip under UHV conditions in a way that a nonlinear I-V curve is obtained which shows only little local variation. Therefore one should be able to map directly the local electric field.

We believe that the coupling of laser light into the tunneling gap by means of surface plasmons opens up a new field of experiments. Since the laser intensity used in our experiment is several orders of magnitude smaller than in other experiments combining laser radiation and STM, thermal effects are negligible. The range of available wavelength covers the infrared and the visible spectrum. The experimental arrangement is therefore ideally suited to investigate e.g. adsorbed atoms, molecules etc. in the presence of an electric field of optical frequency by STM.

Acknowledgement

I would like to thank my co-workers C. Baur, A. Esslinger, U. Graf, B. Koslowski and P. Kürz. I gratefully acknowledge very stimulating discussions with U. Albrecht, J. Boneberg, P. Leiderer and K. Dransfeld. Part of this work is supported by the Deutsche Forschungsgemeinschaft (SFB 306).

References

1 S. Park, C.F. Quate, Appl. Phys. Lett. **48**, 112 (1986)
E. Stoll, O. Marti, Surf. Sci. **181**, 222 (1987)
2 D.W. Abraham, C.C. Williams, H.K. Wickramasinghe, Appl. Phys. Lett. **53**, 1503 (1988)
3 R. Möller, A. Esslinger, B. Koslowski, J. Vac. Sci. Technol. **A8**,590(1990)
4 R. Möller, A. Esslinger, B. Koslowski, Appl. Phys. Lett. **55**,2360(1989)
5 H.E. Farnsworth, R.P. Winch, Phys. Rev., **58**,812 (1940)
6 P. Muralt, D.W. Pohl, Appl. Phys. Lett. **48**,514(1986)
P. Muralt, H. Meier, D.W. Pohl, H.W.M. Salemink, Appl. Phys. Lett. 50,1352(1987)
7 J. Kirtley, S. Washburn, M.J. Brady, Phys. Rev. Lett. **60**,1546(1988)
J.P. Pelz, R.H. Koch, Rev. Sci. Instrum. **60**,301(1989)
8 J.M.R. Weaver, L.M. Walpita, H.K. Wickramasinghe, Nature **342**,783 (1989)
9 H.K. Wickramasinghe, IBM Europe Institute, Garmisch Partenkirchen, August 14-18, 1989
10 N.M.Amer, A.Skumanich, D.Ripple, Appl.Phys.Lett.**49**,137(1986)
11 P.H.Cutler, T.E.Feuchtwang, T.T.Tsong, H.Nguyen, A.A.Lucas, Phys.Rev.**B35**,7774(1987)
12 W.Krieger, T.Suzuki, M.Völcker, H.Walther, Phys.Rev.**B41**,10229(1990)
13 L.Arnold, W.Krieger, H.Walther, Appl.Phys.Lett.51,786(1987)
14 R.J. Hamers, K. Markert, Phys. Rev. Lett. **64**,1051(1990)
Y. Kuk, R.S. Becker, P.J. Silvermann,G.P.Kochanski,Phys. Rev. Lett. **65**,456

15 J.K.Gimzewski, J.Sass, R.R.Schlittler, J.Schott, Europhys.Lett.8,435(1989)

16 R.Berndt, A.Baratoff, J.K.Gimzewski, Proceedings of the NATO Meeting held in Erice, Italy, April 17-29, 1989(Kluwer Academic Publishers)

17 D.L. Abraham, A. Veider, Ch. Schönenberger, H.P. Meier, D.J. Arent, and S.F. Alvarado, Appl. Phys. Lett.56,1564(1990)

18 see e.g. H.Metiu, Progr.Surf.Sci.17,153(1984), M.Moskovits, Rev.Mod.Phys.57,783(1985), A.Otto et al., Surf.Sci.210, 363(1989)

19 R.Möller, A.Esslinger, M.Rauscher, J.Vac.Sci.Technol.A8,434(1990)

20 E.Kretschmann, H.Raether, Z.Naturforsch. A23,2135(1968)

21 A.Tagliacozzo, E. Tosatti, "Tunneling with coupling to a surface phonon or plasmon", Proceedings of the Adriatico Research Conference on "Scanning Tunneling Microscopy", July 28-31, 1987

AN STM STUDY OF THE OXYGENATION OF SILICON

M. E. Welland and R. B. Leane
Cambridge University Engineering Department
Trumpington St.
Cambridge CB2 1PZ UK

Abstract

We have made a detailed atomic study of the early stages of oxidation of silicon at room temperature. On the reconstructed 111 surface the reaction occurs rapidly with a sticking coefficient of ~0.02. Three types of defect or reacted site are found on the surface; the number density of the two commonest defects increasing linearly with oxygen dose. The commoner of the two defect types exhibits image contrast consistent with a depletion in electron density from empty (anti-bonding) to filled (bonding) states which we believe is due to the reaction of the oxygen in the layer immediately below the adatoms. Tunnelling spectroscopy at such sites on the surface shows the quenching of the adatom states and the appearance of a peak in the density of states at -0.8 eV. The other common defect appears in the filled state image exhibiting weaker contrast than the first type and indicating an enhancement in the filled state density. Both defect types preferentially react in the faulted half of the unit cell with an 88% probability. The third type of defect, with only a weak dependence on oxygen dose, occurs at the corner of the unit cell and produces a large change in the electron barrier height consistent with uncompensated charge or an electron trap.

Introduction

The commercial success of the modern micro-electronics industry is largely attributable to the invention in 1960 of metal-oxide-semiconductor (MOS) devices, and therefore on the excellent physical and electronic characteristics of the Si-SiO_2 interface. This structure is not only used as an integral part of MOS devices, but also for the electrical isolation of devices and for surface state passivation. As a result of this success the oxidation of silicon has become one of the most common processes used in today's fabrication of small electronic devices.

Thin oxide (<300Å) characteristics

Both the oxide structure and growth kinetics of SiO_2 begin to change when the oxide thickness (or region of interest) is less than 300Å. This is an increasingly important regime, since devices are already being made with these thin oxides; for example the short channel MOSFETs (Dennard et al., 1979), photovoltaic cells where oxide layers of order 30Å increase the efficiency of the cell, and now in dynamic RAMs where the oxide thickness determines the capacitance of the device.

The thin oxide regime is not as fully understood as the bulk oxide regime, and there is a large amount of evidence to show that the physical and electronic properties of the two are quite different. After the interfacial silicon crystal region, there is a thin layer, approximately 5-10Å thick, of suboxides comprising SiO_4, Si_2O_3, Si_3O_2 and Si_4O tetrahedra (Grunthaner and Grunthaner, 1979). The boundary between the silicon and this mixed oxide layer is extremely abrupt; there is a plane of silicon atoms that is bonded on one side to more silicon atoms, and on the other side to oxygen atoms (Wagner et al., 1985). This first oxide layer is then followed by a region, usually less than 50Å thick, which is strained and largely consists of SiO_2, but with some Si_2O_3 intermixed (Hecht et al., 1985). The remainder of the oxide then consists of stoichiometric amorphous SiO_2 that is strain free.

Thin oxide growth

The growth of thin oxides and their resulting electronic and physical properties are heavily influenced by factors such as the sample cleanliness and the presence of impurities in the sample or oxidising atmosphere (e.g. dopants or water). Unfortunately, strain at the interface tends to harbour these impurities and defects, accentuating the problems of growing reliable and uniform films. One also finds that the linear oxidation regime extrapolates to an initial oxide thickness of typically 200Å, implying that there is an initial unexpected rapid growth of oxide (Massoud et al., 1985), further complicating the understanding of thin oxide growth.

The initial growth of thin oxides is set by the surface reaction rate but as soon as the oxide becomes three-dimensional, strain effects at the surface become significant. A build up of charge at the interface may also affect the initial stages of oxidation. This is possibly due to the production of O^- ions by tunnelling electrons and oxide charges. These O^- ions would have a greater mobility than O_2 molecules, thereby enhancing the oxide growth rate (Hu, 1983). It is also known that there is a build up of fixed charge in the oxide due to the presence of impurity ions such as dopants. This charge induces band bending and reduces the hole density, which in turn will affect the number of Si-Si broken bonds present, and therefore the oxidation rate (Schafer and Lyon, 1985).

Current techniques for growing thin oxides are varied, and the method used can have significant effects upon the quality of the final oxide layer. Reducing the oxygen partial pressure enables oxides in the range 30-150Å thick to be made with high breakdown fields (~10MV/cm) that are highly uniform and homogeneous (Massoud et al., 1985). Similar quality oxides can also be made by using a low temperature (750°C) high pressure (10atm) steam oxidation method.

Structure of the Si (111) - (7x7) reconstruction

The bulk terminated Si (111) surface has one dangling bond per silicon surface atom. The effect of the (7x7) reconstruction is to significantly reduce the number of dangling bonds, and therefore the energy associated with the sudden termination of the bulk lattice. The unit cell in the dimer-adatom-stacking (DAS) model (figure 1a) contains only 19 dangling bonds, compared to the 49 that would be present on an equivalent area if the surface were not reconstructed (Takayanagi et al., 1985). The unit cell consists of two triangular sub units, and each sub unit is surrounded by nine silicon dimers. The top layer of the unit cell consists of 12 adatoms, which locally form a 2x2 structure. These two sub units are stacked slightly differently to each other with respect to the layer of silicon atoms directly below, and this gives rise to a very slight difference in height between the two halves. The two halves are then denoted the faulted and unfaulted halves of the unit cell. Finally there is a vacancy at each of the four corners making a total of 1 vacancy per unit cell.

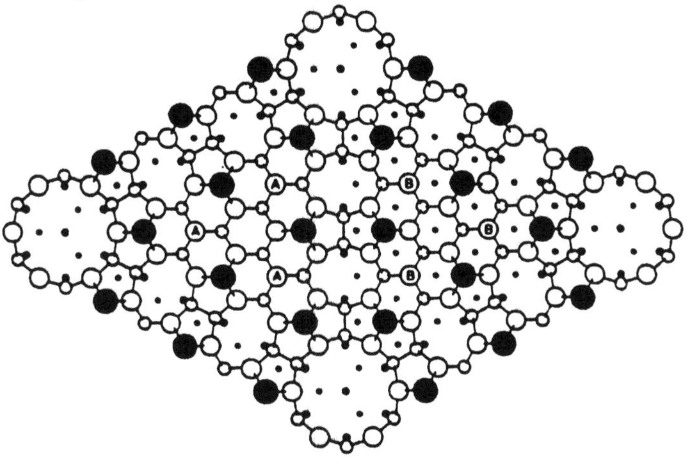

1a. The top view of the unit cell has atoms at greater depth indicated by circles of decreasing diameter. The 12 adatoms that are imaged by STM are represented by solid black circles. The 6 rest atoms sites are marked A and B, and the four vacancies are seen at the corners of the unit cell.

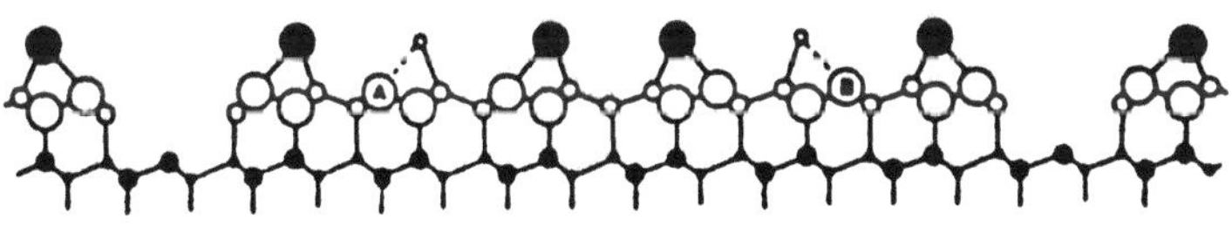

1b. The side view represents a slice taken along the long diagonal of the unit cell. Atoms behind this diagonal are represented by smaller circles. Between the two vacancies are 6 adatoms (4 large and 2 small circles). 2 rest atoms are marked A and B. The different stacking sequence relative to the bulk lattice of the two halves of the unit cell can also be seen.

Figure 1: Schematic of the dimer-adatom-stacking (DAS) model of the Si(111)-(7x7) surface (Takayanagi et al., 1985). The faulted half of the unit cell is on the left.

The 19 dangling bonds are located over three different sites within the unit cell. 12 of them are located over the 12 adatom sites that form the top layer of the unit cell. A further 6 lie over the 6 rest atoms that lie between the adatoms and the bulk, and are marked A and B in figure 1. The last is located in the vacancy at the corner of the unit cell. It is the dangling bonds that are most likely to react with colliding molecules so that the Si atoms associated with the bonds can complete their electron shells. Due to the chemically different locations of the dangling bonds within the unit cell, it is quite likely that they will demonstrate different reaction rates. This possible difference of reactivity across the unit cell should also be enhanced at low temperatures when the difference in activation energies will have a more significant effect.

Since the STM is usually sensitive to the top atomic layer only, STM images tend to reveal solely the 12 adatoms. When imaging empty surface states (i.e. at negative tip bias), the unit cell appears to be completely symmetrical (Figure 2a), but when imaging filled states (i.e. at positive tip bias) there is a slight asymmetry between the faulted and unfaulted halves of the unit cell (Figure 2b). As predicted by the DAS model, the faulted half appears slightly higher, as do the six corner atoms.

Experimental procedure

1. Sample preparation

P-type Si (111) with a resistivity of 0.1-1.0Ωcm from device quality wafers was used in all the experiments that follow. Care was taken to avoid bringing the silicon into contact with metals such as stainless steel since these leave small impurity concentrations on the surface that can reorder, or even completely prevent the formation of, the surface reconstruction. Samples were then transferred through the sample exchange mechanism into UHV at a pressure of 4.10^{-10}torr. The sample was resistively heated using a DC current limited power supply, first to outgas the sample holder, and then larger currents were used for final sample cleaning and preparation. During this preparation the temperature of the sample was monitored with a digital optical pyrometer, correctly adjusted for the emissivity of the surface and the transmissivity of the quartz windows of the UHV system. Great care was exercised in order to ensure that the background pressure never rose above 1.10^{-9} torr while outgassing the sample holder.

The reconstruction was then obtained by flashing the sample to 1250°C for 30s, again ensuring that the pressure stayed below 10^{-9}torr. This effectively cleaned the surface either by boiling off the impurities as in the case of SiO_2, or by dissolving them into the bulk crystal in the case of carbon. The sample was then cooled over a few seconds to 850-900°C where it was left to anneal for 3-5 minutes. After this the sample was very slowly cooled to room temperature by gradually reducing the current over a period of approximately 10 minutes. Auger spectroscopy was then used to ensure that the surface was free of contaminants such as oxygen, carbon and nickel.

This preparation technique reliably produces surfaces on which there are many large atomically flat regions several hundred Angstroms square in size , separated by long straight atomic steps that are 2.3Å high, or multiples thereof . These large flat terraces are then usually covered by the (7x7) surface reconstruction (Figure 2).

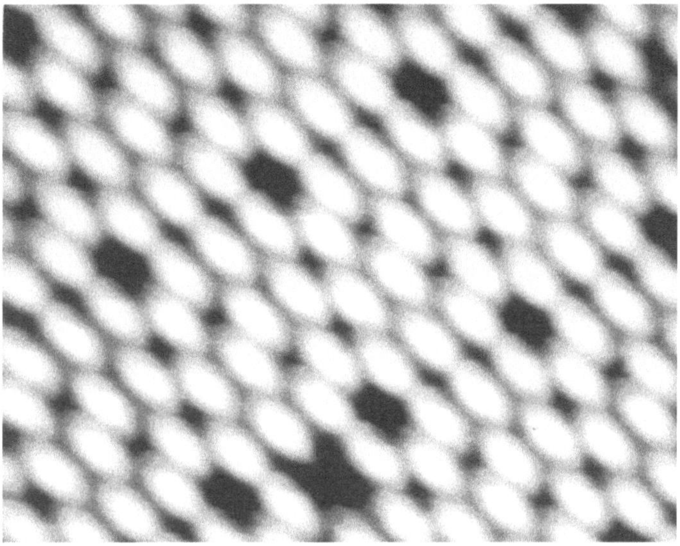

Figure 2a: Si(111)-7x7 empty states. V_{tip} -1.5V. Area ~75Å^2.

Figure 2b: Si(111)-7x7 filled states. V_{tip} 1.5V. Area ~75Å^2.

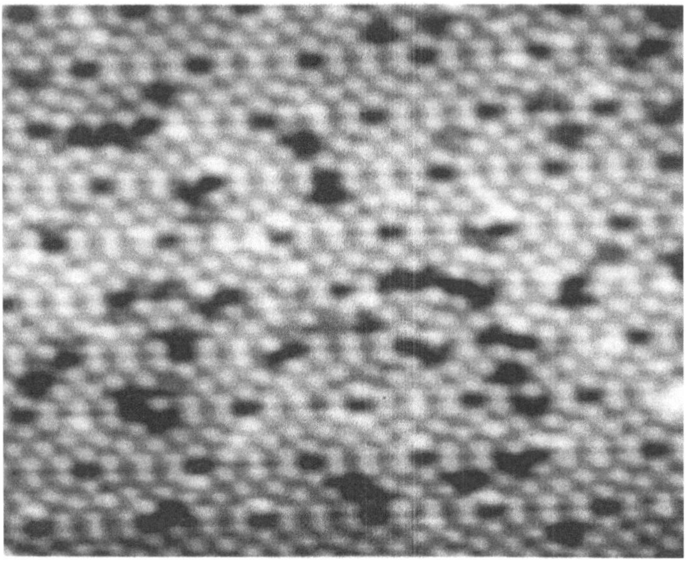

Figure 3a: After 0.25 LM oxygenation. V_{tip} -1.7V. Area 150Å^2.

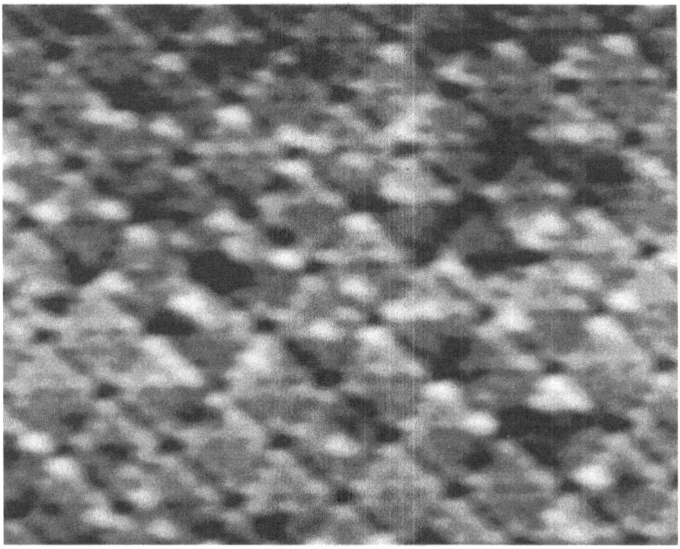

Figure 3b: Identical area to figure 3a except V_{tip} +1.7V.

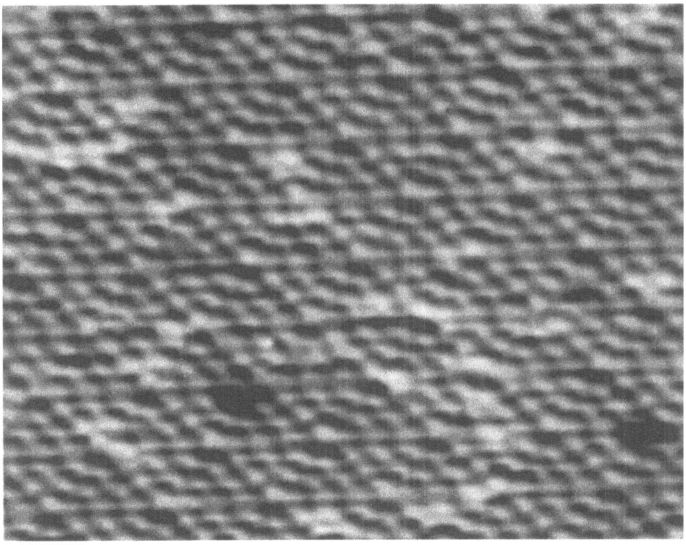

Figure 3c: Identical area to figure 3a but a barrier height image.

2. Silicon oxygenation

Ultra high purity (UHP) oxygen was admitted into the UHV system in controllable quantities via a high precision leak valve. The oxygen partial pressure was measured with a mass spectrometer that had been calibrated correctly for the oxygen cracking pattern. During these experiments the ion gauge was turned off since the presence of hot filaments in the vacuum system may produce dissociated, ionised or excited oxygen species. Pt-Ir filaments were fitted to the mass spectrometer head to minimise the production of CO due to the reaction of oxygen with hot carbon often found in tungsten filaments. The effect of the mass spec. filament in producing excited species is believed to be negligible (Gupta et al., 1989).

The following two methods for examining the dynamic oxygenation of the clean silicon surface were used:

1) Between taking images the tip is retracted electronically so that it is about 0.5μm away from the surface. A short burst of oxygen at a high partial pressure is then admitted into the UHV allowing say a 0.25LM coverage (1LM=10^{-6}torr for 1 second). Imaging is then resumed and the whole process is repeated several times.

2) Allow a constant low oxygen partial pressure of the order of 10^{-9} torr into the STM chamber. The same area of the surface is then constantly imaged over the next hour or so until a good coverage is formed.

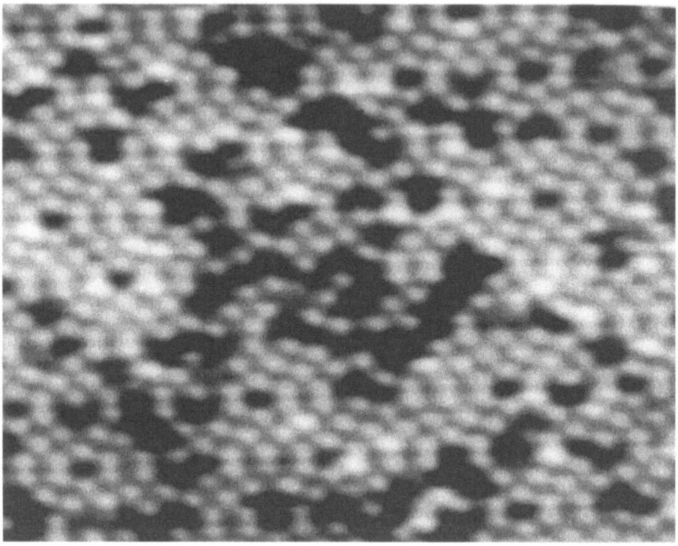

Figure 4a: After 1.0 LM oxygenation. V_{tip} -1.7V. Area 150Å^2.

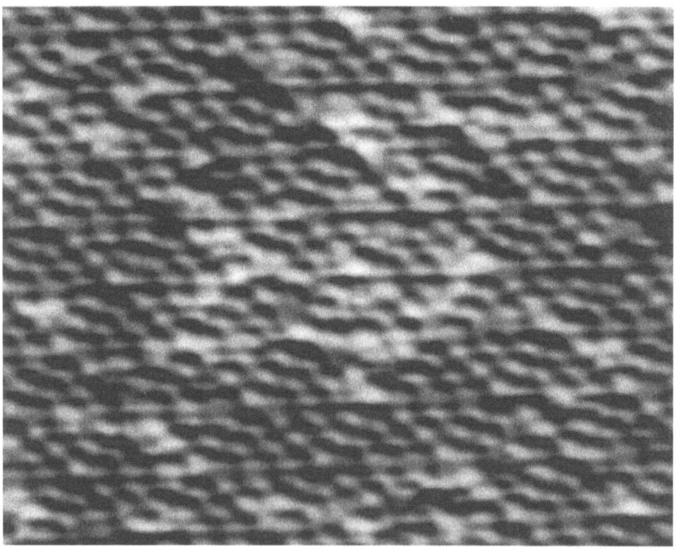

Figure 4b: Identical area to figure 4a except barrier height image.

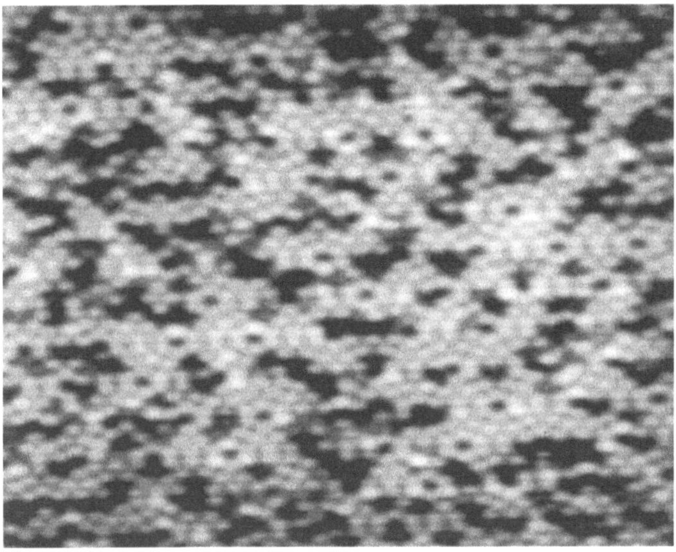

Figure 5a: After 2.0 LM oxygenation. V_{tip} -1.7V. Area 220Å^2.

Figure 5b: Identical area to figure 5a but V_{tip} +1.7V.

At first sight one might expect the results from these two oxygenation methods to be effectively identical. However, there is a very high electric field of order $10^9\,Vm^{-1}$ between the tip and surface, and this, together with possible geometric shielding effects of the tip, may have a significant effect upon the end result. Indeed, this high electric field has already been used in carefully chosen atmospheres in order to deposit extremely high resolution structures onto various surfaces (Ehrichs et al., 1988). In both cases, images were taken at positive and negative tip voltages in order to image both the filled and empty surface states respectively. In addition, barrier height images were taken and plots of tunnelling current against tip voltage at constant gap were also measured.

3. Comparison of clean and oxygenated surface

The effects of exposing the clean surface to 0.25LM of oxygen can be seen in figure 3. This figures contain three images, one of the empty states (figure 3a), one of the filled states (figure 3b), and one of the work function (figure 3c). The comparatively small oxygen exposure is seen to have significantly degraded the surface, and this is clearly imaged by the STM. The most striking change is to the empty states image (figure 3a), in which a large number of hole defects appear to have replaced individual adatoms of the reconstruction. There is no immediately obvious grouping or preferential positioning of these defects. Comparison of the empty states image with the corresponding filled states (figure 3b) image shows that the hole defects just mentioned are not as deep in the filled states image. This is not immediately obvious in the photographs, firstly because the normalisation procedure makes the hole black in both images, and secondly because the process of photographing the images tends to introduce additional contrast, further masking the effect.

Examination of the filled states image (figure 3b) shows not only the hole defects, but also a second type of defect that appears as a white 'blob' on the surface (hereafter 'raised defect'). Like the hole defects, these are at first sight randomly distributed over the adatom sites of the reconstruction. Comparison with figure 3a shows that these protruding defects are present in the empty states images, but they are much less prominent. In general the positions of the hole and raised defects are not correlated, although there are occasions when they overlap with each other when comparing empty and filled state topographic images of the same area.

Examination of the work function image (Figure 3c) shows more changes to the surface that have been effected by the oxygenation process. The atomic reconstruction is clearly visible in the barrier height image, but also there are a number of white areas representing increased barrier height. By comparison of figures 3a and 3c it is evident that these areas of higher barrier height correspond to hole defects in the empty states image. We can also see that the raised defects correspond to regions of decreased barrier height. As in the topographic images, the perturbation to the barrier height is localised over what were individual adatom sites of the surface reconstruction.

Also in figure 3c, we see that there are occasional very deep depressions of the barrier height (black in the figure); much deeper than the perturbation discussed in the previous paragraph. This depression appears only in the corners of the unit cell immediately adjacent to the corner hole. Comparison of the barrier height images with empty and filled state topographic images shows that there is no change to the topographic images where these depressed barrier height sites lie. This type of defect behaves as an electron trap (Welland and Koch, 1986) when the tip is held stationary over it and may be associated with such traps seen on the silicon surface previously. The large change in barrier height then corresponds to the trap being in a charged state; the charge behaving as a coulombic block to tunnelling electrons.

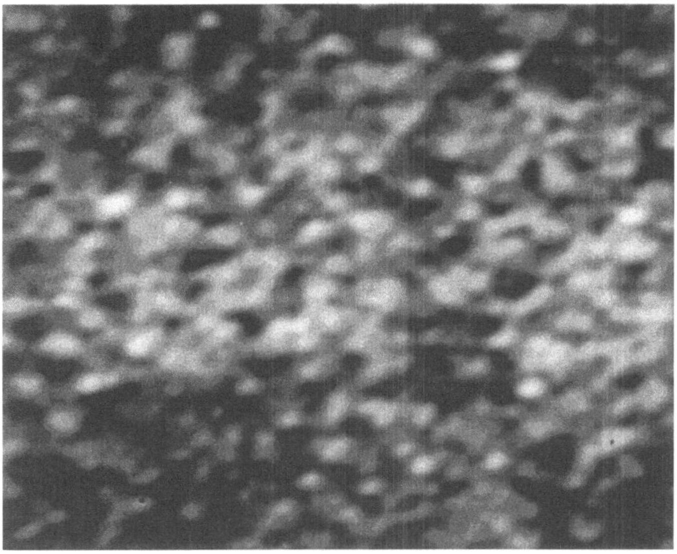

Figure 6: After 4.0 LM oxygenation. V_{tip} -1.7V. Area 150Å^2.

4. A sequence of images taken after successive oxygen bursts

A selection of images taken from a sequence while admitting successive bursts of oxygen into the system is shown in figures 4 to 6. During this sequence of images, the gradual erosion of the surface by the oxygen can be clearly seen in the empty states images, with an approximately linear increase in defects with oxygen exposure. Similarly, there is an increase with oxygen exposure of the number of raised defects in the filled states images. Eventually after about 4LM coverage, it becomes very difficult to obtain good images of the surface at all, with atomic resolution becoming impossible over most of the surface (figure 6). This loss of resolution could be the result of two effects. Firstly the oxide layer will be somewhat amorphous and this generally produces a surface density of states that varies only slightly with respect to atomic positions. Secondly, the oxide layer may have both ionic charge Q_m and oxide fixed charge Q_f in it. The effect of these electrostatic charges may well be greater than the small atomic variations that are present, and therefore overwhelm any possibility of atomic resolution that may be present.

Spectroscopic results

The surface states of the clean (7x7) surface have been studied using techniques such as ultraviolet photoemission spectroscopy (UPS), inverse photoemission spectroscopy (IPS) (Himpsel and Fauster, 1984) and STM (Hamers et al., 1987). Results from UPS and IPS are shown in figure 7a, and they show five prominent peaks. These peaks lie at energies of -1.6, -0.85, -0.1, +0.5 and +1.6eV relative to the Fermi energy. The peak at -1.6eV corresponds to backbond states that are localised between the Si adatoms. At -0.85eV are the six filled dangling bonds of the 6 rest atom sites. Adatom states lie at -0.1eV, and these states are stronger on the faulted half of the unit cell than on the unfaulted half, giving rise to the characteristic asymmetry of the STM images of filled states. The empty states at 0.5eV lie over the adatoms, and the states at 1.6eV lie between the adatoms.

To probe the surface states of the oxygenated surface, a number of I-V characteristics (hereafter 'IVs') were taken over the (7x7) reconstruction at various points during the oxygenation of the surface. A large number of IVs were taken and in this way groups of characteristic IVs were formed. A measure of the local density of states is then obtained from these IVs by numerically calculating $\partial \ln I/\partial \ln V$.

IVs taken over the clean surface yielded a density of states with peaks in very good agreement with those of the UPS and IPS data (figure 7b). A series of IVs were then taken over the surface after it had been exposed to 1LM of oxygen, and the surface had a significant coverage of oxygen defects. A group of IVs is shown in figure 8a together with their average. On calculating $\partial \mathrm{Ln} I/\partial \mathrm{Ln} V$, it was found that several of them showed a LDOS very similar to the LDOS observed on the clean silicon surface (figure 8b), while some of them were quite different (figure 8c). The most noticeable change is the elimination of the peak at ~0.8V; i.e. the states corresponding to the rest atom dangling bonds. (Remember that a state at tip voltage V_{tip} corresponds to a state at energy $-eV_{tip}$ relative to the Fermi energy).

At a slightly higher oxygen coverage the basic form of the IVs begins to change quite significantly. Figure 9a shows a group of IVs taken on a surface that has been exposed to 1.4LM of oxygen. Whereas the previous IVs showed similar currents at +2V and -2V tip bias, these IVs show a strong asymmetry, with the tunnelling current at +2V averaging to only ~1.1nA compared to ~2.5nA at -2V tip bias. This set of IVs is that again forms two sets, in which one of the sets shows some surface states at approximately -0.8V (i.e. at an energy of 0.8eV relative to the Fermi energy). This surface state is enhanced in the $\partial \mathrm{Ln} I/\partial \mathrm{Ln} V$ plot in figure 9b.

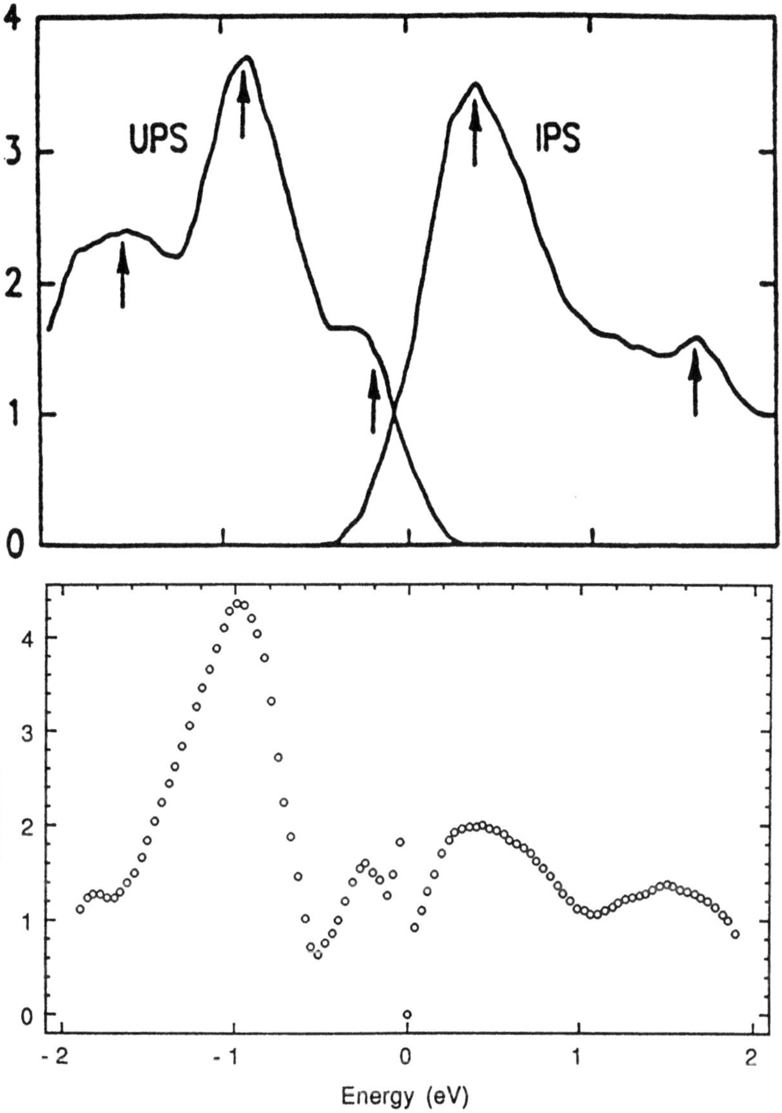

Figure 7a and b: Surface density of states measurements of Si (111)-(7x7) reconstruction measured by photoemission and inverse photoemission (top) (Himpsel et al 1984), and by STM (bottom).

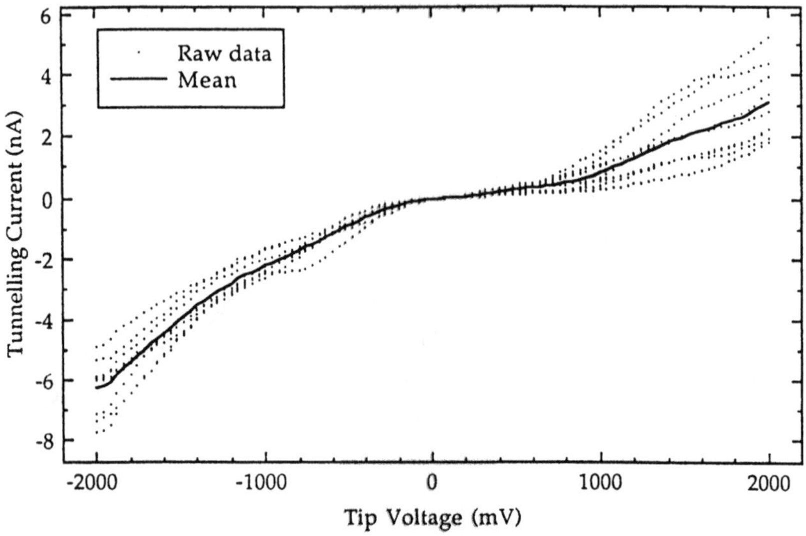

Figure 8a: I-V characteristics taken on the Si(111)-(7x7) surface after 1LM exposure to oxygen.

Another group of IVs, taken on a surface that had been exposed to 2LM of oxygen, is shown in figure 10a. The characteristics are now quite different at positive and negative tip bias, and the tunnel junction is now showing a strong rectifying type behaviour. It is this behaviour that is responsible for the difficulty in obtaining good images of the surface at positive tip voltage, and necessitated an increase in the tip voltage from 1.5V to >2.5V. This rectification is characteristic of a typical MIS (metal-insulator-semiconductor) diode. The reason for the sudden change from a non rectifying behaviour can be attributed to the reduction of the surface states of the silicon. This in turn stops the pinning of the Fermi level, and the surface can no longer have the 'metallic' character of the clean surface. This enables band bending and therefore the usual accumulation, depletion and inversion regions characteristic of MIS systems. The plot of $\partial \ln I/\partial \ln V$ shown in figure 10b shows a similar LDOS peak at -0.8V to the case of 1.4LM oxygen exposure. Since the clean surface has no LDOS peak at -0.8V, this suggests that there is a new state arising from the oxygenation.

IVs taken on a surface exposed to 7LM of oxygen showed no significant change to previous IVs.

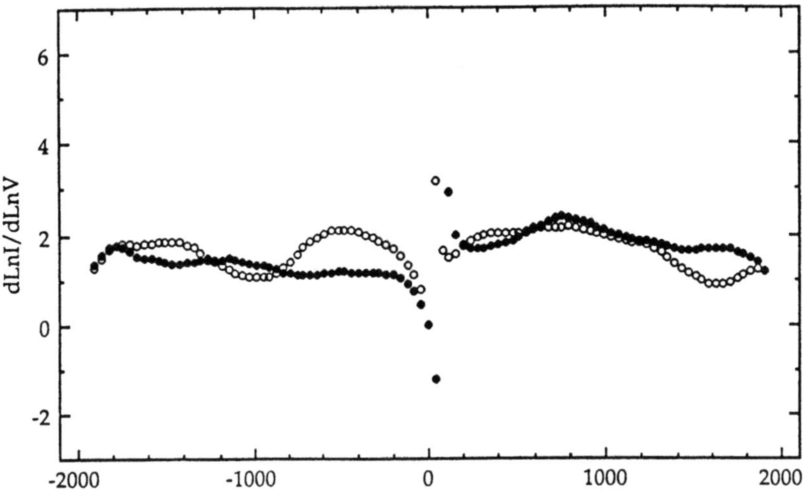

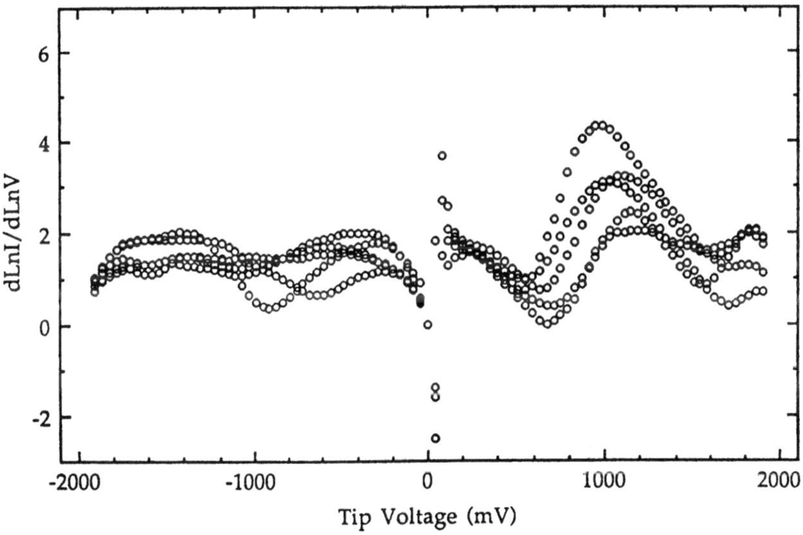

Figure 8b and c: Density of states peaks on Si(111)-(7x7) surface after exposure to 1LM of oxygen. Some plots show strong peaks at 0.8V corresponding to rest atom dangling bond states (bottom), while others do not (top).

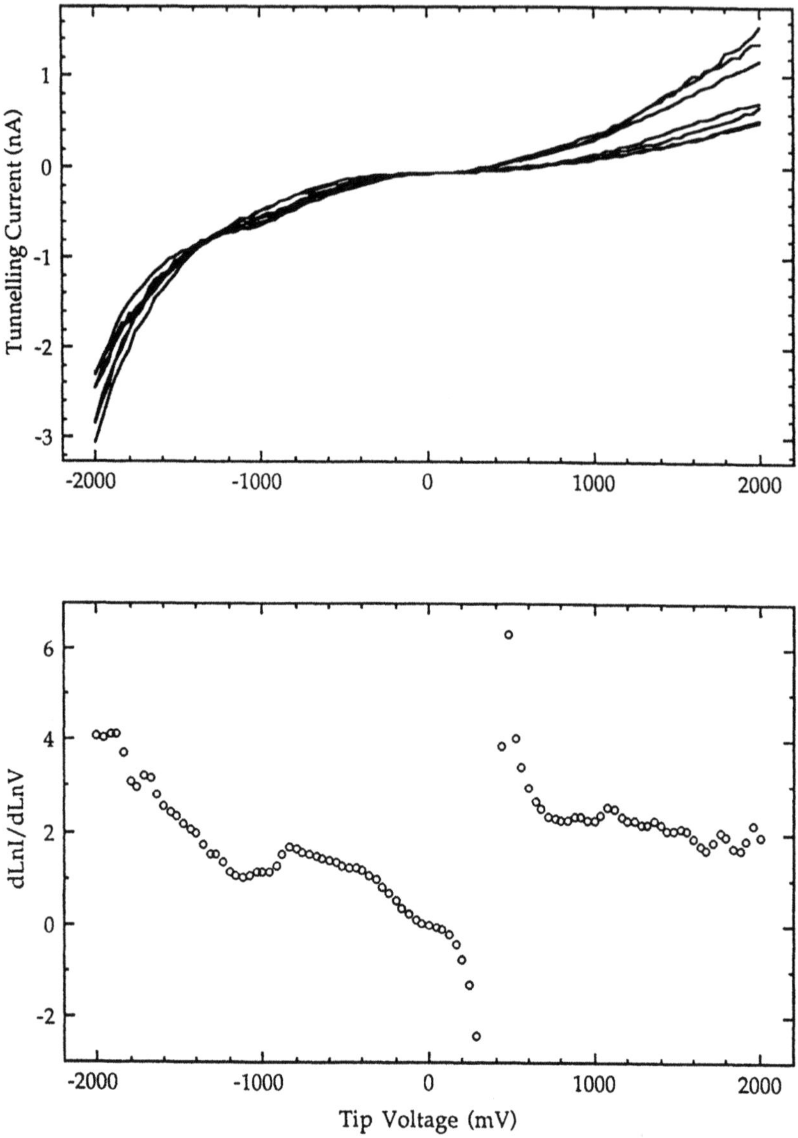

Figure 9a and b: I-V characteristics taken on the Si(111)-(7x7) surface after exposure to 1.4LM of oxygen (top). The normalised derivative plot (bottom) shows the density of states peaks, with a new peak developing at an energy corresponding to V_{tip} ~-0.8eV.

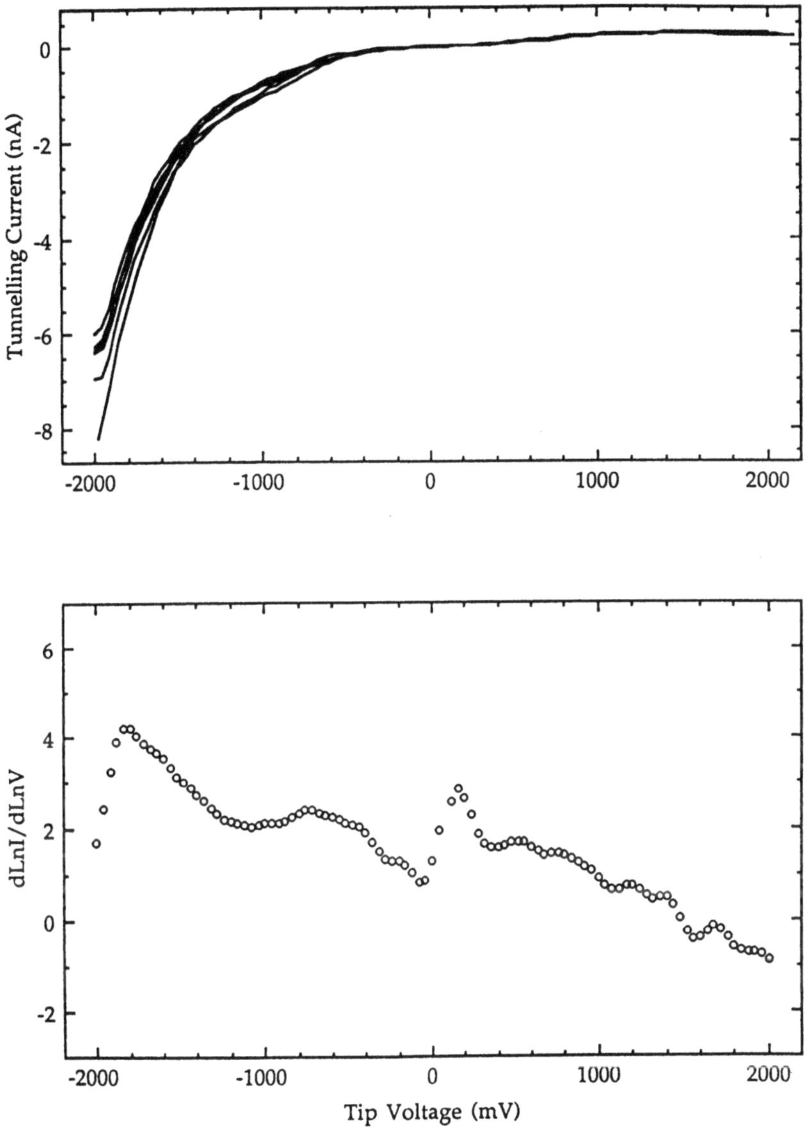

Figure 10a and b: I-V characteristics taken on the Si(111)-(7x7) surface after exposure to 2LM of oxygen (top). The normalised derivative plot (bottom) shows a density of states peak at an energy corresponding to V_{tip} ~-0.8eV.

Discussion

1. Imaging Results

Closer inspection of figures 3 to 6 shows a variety of interesting features. During the course of the experiment (~3 hours) there is no observation of mobility of any of the defects, indicating that the defects are the result of a reaction rather than the mere physisorption onto the surface.

The apparent depth of the hole defects in the empty states images is approximately 1.0Å. This was measured over an extended reaction region in order to prevent errors from the finite tip shape brushing on the edges of the hole. These holes are also accompanied by a 12% increase in the local barrier height as seen in the relevant barrier height images (see figures 3c and 4b). The raised defects found in the filled states images sit ~0.6Å above the adatoms, and have a 15% decrease in the local barrier height associated with them. The lateral dimensions of the defects are both of similar dimensions to the silicon adatoms themselves.

When interpreting topographic images, it is important to realise that changes in the topography at a reaction site can be the result of three different effects:

1) A genuine change in topography due to the removal or insertion of an atom.

2) While scanning in the constant current mode, the product of the local barrier height ϕ and the square of the vacuum gap s must be constant (ϕs^2=constant). Thus a change in the local barrier height will also manifest itself as a change in surface topography.

3) A change in the chemical nature of the site may alter the LDOS. The transfer Hamiltonian formalism of tunnelling theory shows how the tunnelling current is proportional to the LDOS. Thus a change of the LDOS may affect the topographic image quite independently of the other two effects (Tersoff and Hamann, 1985).

It is possible to estimate (2) if we know the fractional change in the local barrier height together with the size of the vacuum gap. If we assume that the vacuum gap width is about 6Å (Tersoff and Hamann, 1985), then the real depth of the hole, taking into account the 12% change in barrier height and the fact that the product ϕs^2 is constant, is probably nearer 0.7Å. This is a further indication that the features are genuine holes, as opposed to bumps with a large perturbation of the local barrier height. It should be realised that the choice of a 6Å tunnelling gap is somewhat arbitrary, and that if the gap were above 18Å then both of these features would in fact be inverted. However, a gap as large as 18Å is unlikely due to the low tip voltages used and band bending effects. The heights of both types of defect as they appear in filled and empty states images is shown are table 1 together with a correction for the measured change in barrier height.

	empty states	filled states
hole depth	1.0A	0.3A
corrected depth	0.7A	~0A
raised defect height	0.3A	0.6A
corrected height	~0A	0.3A

Table 1: Relative sizes of features on oxygenated silicon. The corrected depth/height refers to the height after a correction for the barrier height change of +12% and -15% for the hole and raised defects respectively. A tunnelling gap of 6Å was used for this calculation.

From the barrier height corrected figures, we see that the holes in the empty states images, and the raised defects in the filled states images, must be the effects of either a perturbation of the LDOS or the topography. The true origin of the defects is found by considering the ways in which the oxygen can react with the silicon.

The first type of defect (the hole defects that are most prominent in the empty states images) could be the result of an oxygen atom burying itself between the top layer adatoms and the second layer rest atoms. The oxygen would then be bonded directly to the dangling bond of the rest atom. The electronegative nature of the oxygen atom would then draw the local anti-bonding states of the silicon adatom (i.e. the dangling bond) down toward itself in an attempt to satisfy the oxygen's own unoccupied dangling bond. This depletion of the LDOS over the adatom would then explain why this type of defect appears as a missing adatom in the empty states topograph.

The second type of defect (the raised defects that are in the filled states images), could be the result of the reaction of an oxygen atom with the dangling bond associated with each adatom. The oxygen atom would then sit on top of the reconstruction and one would naturally expect to see it as a protrusion on the surface. Lang's calculations on the effects of different adsorbates upon the tunnelling current show that an electronegative atom will appear as a large bump when imaging the filled states, but also that the size of the bump will decrease, and even become negative, when imaging the empty states (Lang, 1987). Although Lang treated the case of a sulphur atom, a similar effect would be expected for an oxygen atom, and this is exactly what we see in the topographs.

These models suggesting that the oxygen atoms take up the dangling bonds of both the rest atoms and the adatoms is also in good agreement with spectroscopic results to be presented later.

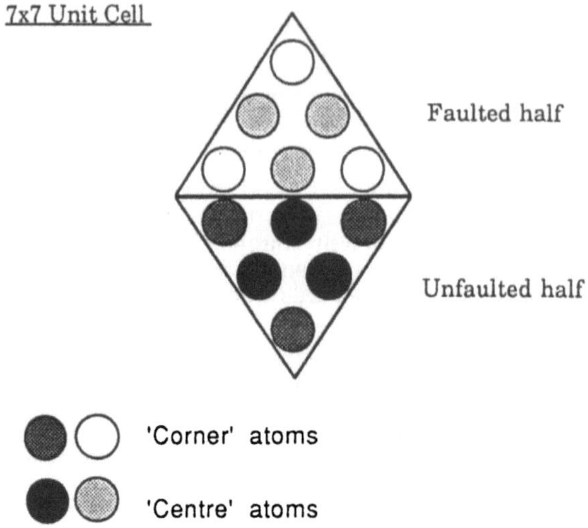

Figure 11: Schematic of the 7x7 unit cell, showing the 4 different reaction sites. 2 of these are at the corners, and 2 are at the edges of the two halves of the unit cell.

2. Defect locations within the unit cell

The hole defects at low exposures appear to be randomly distributed over the silicon surface. As the exposure increases, the defects show a tendency to grow into little groups. As the exposure increases until the whole surface is covered, the images obtained remain very flat with an RMS roughness of <1Å over a 300Å^2 area. This implies that the oxygen defects do not form on top of each other, but that they grow so as to oxygenate the surface uniformly.

To see if there were any preferential reaction sites within the unit cell, the unit cell was divided into chemically different sites, and the number of defects on each site at each coverage was counted. Due to its symmetry, the unit cell was divided into 4 different sites: the corner adatom of each half of the unit cell formed 2 of these sites, and the centre atoms of each half formed the other 2 sites (figure 11).

The number of defects in the unfaulted and faulted halves of the unit cell are plotted as a function of oxygen exposure in figure 12. A linear dependence of the defect density on the oxygen exposure is clearly demonstrated. There is also a slight difference between the reactivity of the corner and centre sites, but this discrepancy is too small to be of any statistical significance. However, there is a considerable difference between the gradients of the two graphs in figure 12, and this shows that the faulted half of the unit cell is ~88% more likely to react with the oxygen than the unfaulted half.

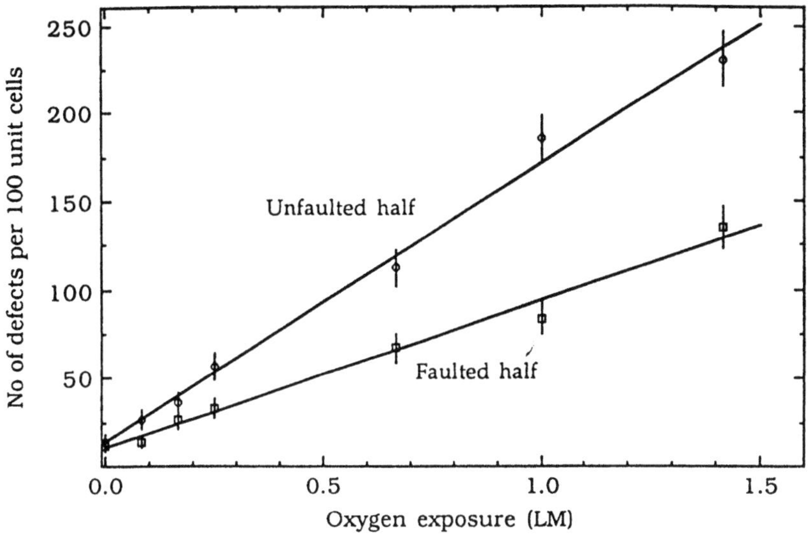

Figure 12: Number of defects found in the faulted and unfaulted halves of the (7x7) unit cell as a function of oxygen exposure.

Forming similar statistics that were reliable for the raised defects on the filled states images was considerably more difficult. Substantial problems were encountered in defining what was and what was not a genuine reaction site. These problems were compounded by the fact that the faulted half of the unit cell is slightly higher than the unfaulted half, and that the six corner atoms are slightly brighter than the six centre atoms. In particular this can make the corner atoms of the faulted half of the unit cell similar in appearance to the reaction sites on the centre atoms of the unfaulted half of the unit cell. Graphs were plotted with a constant subtracted form the data so that they all passed through the origin (figure 13). Although the error margins are rather high for the reasons already stated, the graphs do show a location preference for the reaction sites. Like the hole defects, these defects show a tendency to collect in the faulted half of the unit cell, with a ~200% increase in likelihood. Furthermore, there appears to be a factor of ~2 preference to form over the corner adatoms within each half of the unit cell. A straight line has been fitted to the defect statistics in figure 13. However, unlike the case of the hole defects, it is not conclusive that the density of these adatom defects is linearly dependent upon the oxygen exposure.

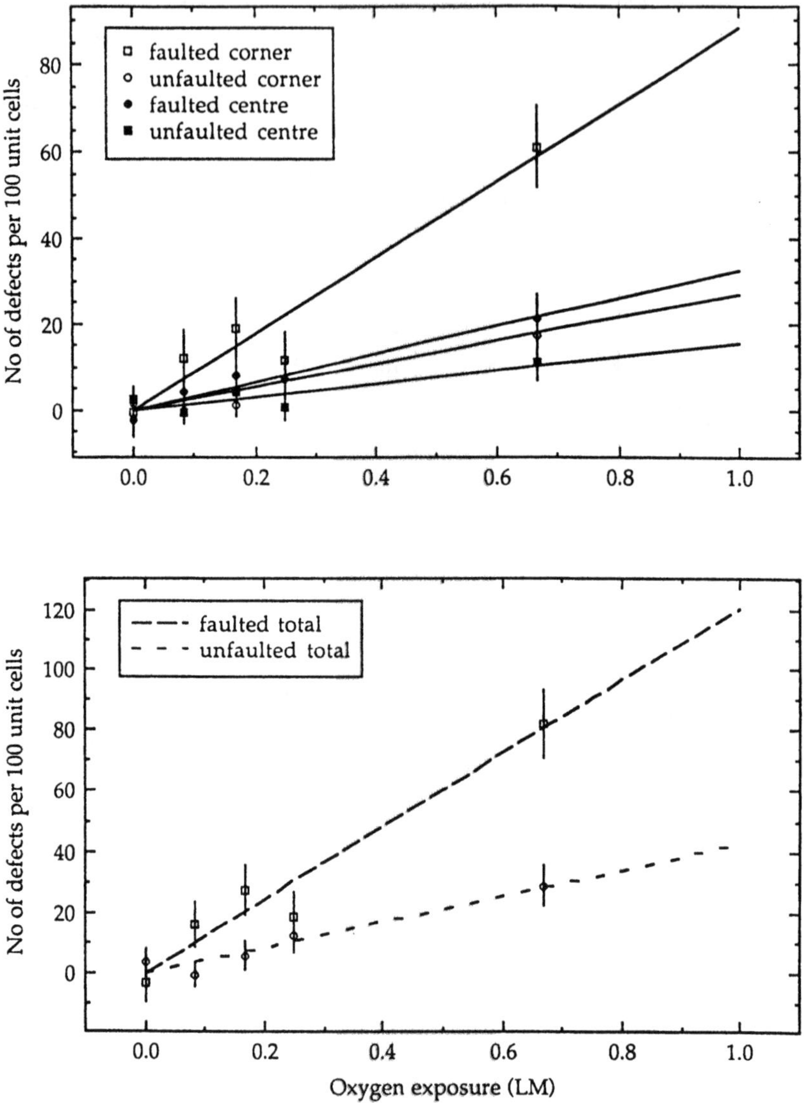

Fig 13: Number of defects found over each reaction site of the (7x7) unit cell as a function of oxygen exposure (top), and the totals for the faulted and unfaulted halves of the unit cell (bottom). Both sets of data have had constants subtracted to ensure that they pass through the origin.

Finally the sticking coefficient can be obtained from the topographic data, and is a quantity of general interest that has been studied by many workers. If all the defects counted are considered to be a result of the reaction with oxygen incident on the surface, then it is a simple matter to calculate that the sticking coefficient is ~0.02. Previous sticking coefficient measurements for O_2 on Si(111)-(7x7) at room temperature have given values varying from $8x10^{-4}$ (Joyce and Neave, 1971) to 0.21 (Kirby and Lichtman, 1974), although most values lie in the range 0.1-0.2.

3. Spectroscopy

The oxygenation of the clean silicon surface is divided into four regimes: the parabolic (1) and linear (2) growth that characterise oxides greater than 300Å thick; an enhanced oxygenation rate for oxides below 300Å (3); and the very fast initial take up of oxygen by the clean silicon surface (4). The last of these processes has been investigated by STM on the clean Si(111)-(7x7) surface at room temperature, and the results of both topographic imaging and spectroscopy have been presented. These results demonstrate five interesting features of the oxygen-Si(111)-(7x7) system, and these are summarised as follows:

1. Two types of defect appeared in the topographic images. The more common defect appeared as a hole in the empty states images and the other as a raised defect in the filled states images.

2) Both defect types were found to favour the faulted half of the unit cell. The defect density of the holes on both halves of the unit cell was linear in oxygen exposure.

3) The LDOS peaks, and in particular the dangling bond peak, obtained from I-V measurements disappeared after 1LM oxygen exposure.

4) After 2LM exposure the junction showed strong rectifying characteristics. A new state appeared corresponding to a tip voltage of ~-0.8V.

It was suggested earlier that the two different types of defect observed had chemically different origins. The topographic images could be explained by an oxygen atom bonding itself to the dangling bond of the rest atom in the second atomic layer of the reconstruction. This oxygen is also able to partially bond itself to a neighbouring adatom, and in this way it forms a Si-O-Si bridging bond. The raised defect on the other hand simply forms a single bond to the dangling bond of the adatom. It then sits on top of the unit cell, and therefore appears as the observed protrusion in the topoghraphic images. These models were used to explain the appearance of the defects in both the filled and empty states images in terms of the affect of the defects on the LDOS.

Using the above models, both of these defects are expected to take up the dangling bonds of the reconstruction. This is in good agreement with the spectroscopy measurements which clearly demonstrate the depletion of the dangling bonds over oxygenated areas of the surface. Other experiments have also suggested that the initial fast take up of oxygen due to reactions with the dangling bonds. AES measurements taken after exposing the surface to 2LM of oxygen show that peaks occur at 84eV and 63/70eV. These peaks correspond to the Si(I) and Si(II) species respectively, where the Si(I) and Si(II) designate oxygen atoms that are bound to one or two silicon atoms respectively (Gupta 1989). Other experiments were performed in which the clean reconstructed surface was first exposed to sub-monolayer coverages of hydrogen. The oxygen sticking coefficient on such surfaces was observed to decrease approximately linearly with increasing hydrogen exposure. It was also found that the saturation exposure, (the exposure at which regime 3 above begins to describe the oxygen take up), set in after a smaller oxygen exposure when the surface had been previously hydrogenated. Since hydrogen is known to passivate the surface dangling bonds, this was found to be clear evidence that the initial adsorption of oxygen on Si(111)-(7x7) requires dangling bonds (Gupta 1989).

These two types of defect have different reactive sticking coefficients. After exposure to 0.67LM of oxygen, the density of the hole defects was nearly double that of the raised defects. This has also been found to be the case in similar experiments in which the Si(111)-(7x7) reconstruction was exposed to sub-monolayer coverages of ammonia (Avouris and Wolkow, 1989). In the ammonia experiments, two types of defect were observed, and were believed to be located on the rest atom and adatom dangling bond sites. They also found that after an ammonia exposure such that half the adatom dangling bonds had reacted, nearly all the rest atom dangling bonds had reacted. However, the reaction of ammonia with silicon is complicated by the possible dissociation of ammonia before reacting, allowing both $Si\text{-}NH_2$ and Si-H products.

To find the cause for the greater reactivity of the rest atoms with the oxygen, we must look more closely at the structure of the unit cell. A proper discussion of the reaction rates would necessitate knowledge of such factors as the reaction activation energies, heats of formation, geometric factors, surface diffusion, energy dissipation and electronic effects of all the possible reaction sites. This is not possible, but since we are merely concerned with relative reaction rates of similar processes in chemically similar environments, it is possible to make some tentative suggestions. Due to the open structure of the (7x7) unit cell, any effect of geometric factors are likely to be small, and would probably favour reaction at the adatom site since it is more exposed to incident oxygen. Due to the similar surroundings of the various possible reaction sites, it is unlikely that surface diffusion and energy dissipation effects are going to significantly affect the relative reaction rates. We therefore look for electronic effects that may produce the observed difference in reaction rates.

The rest atoms are normal three fold coordinated silicon sites with a completely free dangling bond. The adatom sites are directly bonded to the three silicon atoms behind them (figure 14). However, the three silicon atoms behind the adatom are not positioned correctly for the normal tetrahedral structure to form, and there is a certain amount of strain associated with this configuration. This strain pulls the adatoms into the surface and closer to the silicon atoms directly below them. Consequently, the adatom may be able to form a weak fourth bond with the silicon atom directly below it. Evidence for this possible fourth bond has been reported before, where electron energy-loss spectra (EELS) have shown bond vibrations that were explained in terms of a weak bond between the adatom and the silicon atom below (Daum et al., 1987). Thus the reaction of an adatom with an incident oxygen atom would essentially produce an hypervalent five-fold coordinated silicon adatom, whereas reaction with the rest atom would merely involve the saturation of a free dangling bond. Although silicon often forms hypervalent compounds as a result of the empty 3d electron levels, such formations are generally less favourable than the normal four-fold coordinated compounds. Such effects will lead to a difference between the activation energies for reactions of the rest atom and adatom sites. This in turn may at least partially explain the preferential formation of rest atom defects.

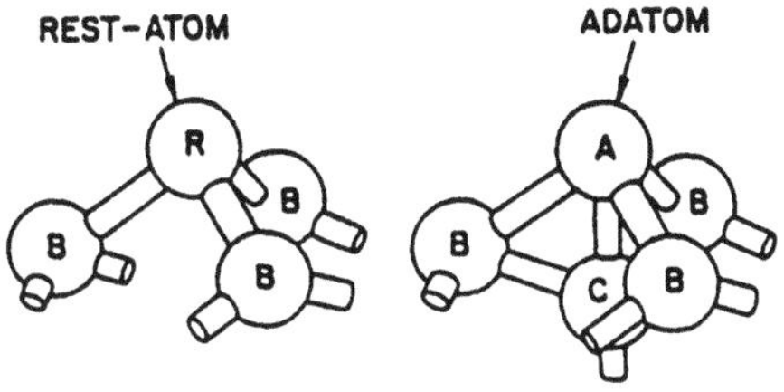

Figure 14: Structure of the rest atom and adatom sites. The rest atom (R) is bonded to 3 silicon atoms (B), leaving a dangling bond pointing upwards. The adatom (A) is also bonded to 3 Si atoms (B), but there is also a weak bond to a silicon atom below it (C).

We now turn our attention to the increased reactivity of the faulted half of the unit cell compared to that of the unfaulted half. This effect has also been observed very recently in STM experiments similar to these (Tokumoto 1990). Tokumoto and co-workers took images of filled states while oxygenating the surface. Their oxygen exposure levels were very high, indicating that they too had problems with the tip hiding the surface from the oxygen atmosphere. They observed a similar increase in reactivity of the faulted half compared to the unfaulted half, but also of the centre adatoms compared to the corner adatoms. However, their results are highly inconclusive. Firstly they refer to the hole defect of the filled states images (the defect we found difficult to see in the filled states images) as the defect that sits on top of the adatoms, and make little reference to the more common rest atom defect. Secondly, the oxygenation was started 17 hours after flashing the sample, and a number of defects were already present before oxygenation was commenced. The strong preferential reactivity of centre adatoms compared to the corner adatoms of the faulted half of the unit cell was the result of counting a total of only 14 defects.

The asymmetry of the unit cell when imaging the filled states suggests that the surface states associated with both the adatoms and rest atoms are stronger in the faulted half of the unit cell than the unfaulted half (Hamers et al., 1987). This is a consequence of the different positions of each half of the unit cell with respect to the bulk lattice. The greater strength of the surface states may facilitate reaction with incident oxygen, and therefore enhance the reactivity of these sites. Hence the greater reactivity of the faulted half of the unit cell is not altogether surprising.

The depletion of the surface states, and the creation of a new state, is also revealed by spectroscopic measurements on the oxygenated surface. The new state appears at an energy of ~0.8eV, and at higher oxygen doses appears to be at a higher energy of ~1eV. However, the I-V characteristics were highly asymmetrical, and were typical of a MIS system. This suggests that band bending effects are significant at the oxygen exposures at which the new state was observed. Such band bending effects must be offset against the measured energy of the state, and for this reason the absolute energy of the state relative to the Fermi energy will be somewhat less than 0.8eV. Increased band bending effects at greater oxygen coverages may also cause the observed increase in tip voltage at which the state is found. The exact energy of the state will depend on the magnitude of the band bending effects and the position of the state within the oxide, and there is insufficient data to determine this.

It is not possible to give a direct cause for this new state. The P_b centres that are known to exist at the Si-SiO_2 interface and have peak densities at ~0.2eV above and below the valence and conduction band edges respectively could provide an explanation. Strain introduced as a result of new bonds can also produce new states. New bonding energies and or the dangling bond of the adatoms defects provide further sources of new states. However, until the exact energy and/or the exact three dimensional spatial position of the state is known, it will be difficult to determine its precise origin.

Acknowledgements

We would like to thank the SERC and the EEC (ESPIRIT basic research action no. 3109 "PROMPT") for financial support. We are also indebted to our partners in the PROMPT program, especially Dr. Carl Sofield and Martin Murrell, for considerable technical input on oxygenation kinetics.

References

Avouris Ph, Wolkow R, *Phys.Rev.B*, 39 #8, 5091 (1989)

Daum W, Ibach H, Müller JE, *Phys.Rev.Lett.*, 59 #14, 1593 (1987)

Dennard RH, Gaensslen FH, Walker EJ, Cook PW, *IEEE Trans. Electron.Dev.*, 26, 325 (1979)

Ehrichs EE, Yoon S, AL de Lozanne, *Appl.Phys.Lett.*, 53, 2287 (1988)

Grunthaner FJ, Grunthaner PJ, Vasquez RP, Lewis BF, Maserjian J, Madhukar A, *Phys.Rev.Lett.*, 43, 1683 (1979)

Gupta P, Mak CH, Coon PA, George SM, *Phys.Rev.B*, 40 #11, 7739 (1989)

Hamers RJ, Tromp RM, Demuth JE, *Surf.Sci.*, 181, 346 (1987)

Hecht MH, Grunthaner PJ, Grunthaner PJ, *Proceedings of the 17th International Conference on the Physics of Semiconductors*, (Springer 1985)

Himpsel FJ, Fauster Th, *J.Vac.Sc.Technol.A*, 2 #2, 815 (1984)

Hu SM, *Appl.Phys.Lett.*, 42, 872 (1983)

Joyce BA, Neave JH, *Surf.Sci.*, 27, 499 (1971)

Kirby RE, Lictman D, *Surf.Sci.*, 41, 447 (1974)

Lang ND, *Phys.Rev.Lett.*, 58, #1, 45, (1987)

Massoud HZ, Plummer JD, Irene EA, *J.Electrochem.Soc.*, 132, 2685 (1985)

Schafer SA, Lyon SA, *Appl.Phys.Lett.*, 47, 154 (1985)

Takayanagi K, Tanishiro T, Takahashi S, *J.Vacuum.Sci.Technol.*, A3, 1502 (1985)

Tersoff J, Hamann DR, *Phys.Rev.B* 31, #2, (1985)

Tokumoto H, Miki K, Murakami H, Bando H, Ono M, Kajimura K, *J. Vac.Sci.Technol.A*, 8 #1, 255 (1990)

Wagner CD, Joshi A, Gulbrandsen L, Deal BE, *J.Vac.Sci.Technol B*, 2 #2, 107 (1984)

Welland ME and Koch RC, *Appl.Phys.Lett.* 48 #11, 726 (1986)

SCANNING NEAR FIELD OPTICAL MICROSCOPY

U. Ch. Fischer

Nanotec GmbH
P.O. Box 6321, D-6330 Wetzlar 21

Abstract:. This article reviews Scanning Near Field Optical Microscopy focusing on the different types of antennas which have been employed as sensors. A proposal for a coaxial tip as a new type of sensor is made in conjunction with a way of its fabrication.

INTRODUCTION

The first proposal of Scanning Near Field Optical Microscopy (SNOM) by O'Keefe (1) is based on a simple idea. A hole of a diameter small compared to the wavelength in a thin metal screen is irradiated with light and light transmitted through the hole is recorded as it is scanned laterally across an object in close proximity. Since the electromagnetic fields penetrating the holes are localised to the dimensions of the hole within the nearfield range extending to a fraction of the diameter of the hole (2,3), a scanned image of the object is obtained at a resolution limited by the diameter of the aperture and not by the wavelength of the light. This concept was first demonstrated in the microwave domain by Ash and Nichols (4), who also extended the concept to a reflection mode, and later in the optical domain by Pohl et al (5) and independently by Lewis et al (6) and by Fischer (7).

More generally, other kinds of an antenna of dimensions small compared to the wavelength may be used instead of an aperture in a metal screen (8,9,10,11).

SNOM is still in an early stage. An extensive review article has recently appeared on this subject (12). Here we discuss shortly results which we obtained with a specific version of SNOM in a reflection mode (12,13), where near field phenomena and their relation to imaging can be clearly shown.

In addition we discuss the different other realisations of SNOM, stressing the aspect of the different kinds of antennas which have been used. The author believes, that further progress in the field will depend strongly on a proper antenna design. In this context the concept of a tapered coaxial cable is introduced as a new type of antenna and a way for its fabrication is outlined. Finally potential applications of SNOM to technology, material sciences and to biology are indicated.

REFLECTION MODE OF SNOM

THE INSTRUMENT

A schematic view of our instrument (13,14) is shown in fig.1. The antennas are holes of 70 - 100 nm in diameter in a 20 - 30 nm thick aluminum film (fig. 1b) or small protrusions of about 70 to 100 nm from a 30 - 40 nm thick film of gold deposited on a glass slide (fig. 1c). A He- Ne laser beam is coupled into the slide such that the antenna is seen in dark field illumination in the light microscope. The object is mounted on a piezoelectric scanning device. The test object is a convex glass surface covered with a pattern of patches only 5 nm thick. The curvature facilitates approach of the antenna while being sufficiently small to make the object appear flat within the scan range. The light scattered from the antenna into the microscope is guided onto a photomultiplier; the signal is stored, together with positional information in an image memory device. Because excitation and observation of the antenna is from the same side, this mode of SNOM is called a reflection mode

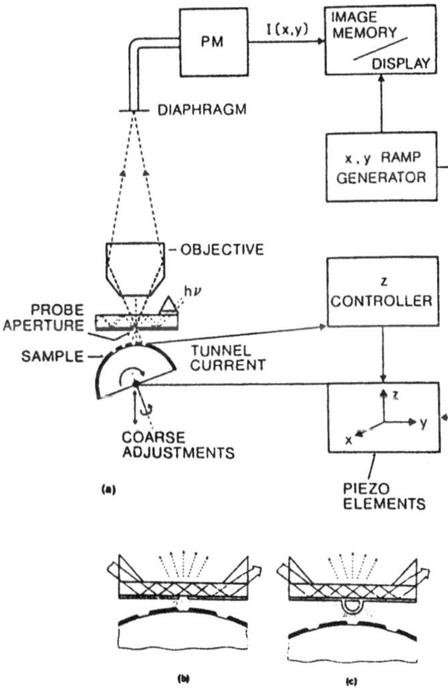

Fig 1.

Experimental scheme for SNOM in reflection.

a)Complete system

b)Arrangement of aperture and sample.

c)Pin and sample arrangement.

Fig. 2 Shows the variation of scattering intensity for the approach of an aperture towards a glas surface covered with a 5 nm thick film of Ta/W (13).

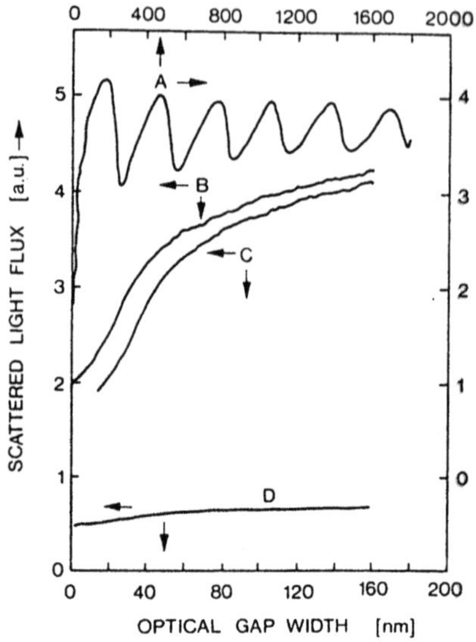

Fig.2

Scattering signal of a 0.09 µm aperture as a function of distance between aperture and sample. Note different scales for curves A and curves B,C, and D. C is a repetition of B, slightly displaced. D. was obtained with the microscope focused away from the aperture. (from (13)).

The approach was stopped in this case with the onset of a tunnel current between the object and the screen containing the aperture. Interference undulations of $\lambda/2$ periodicity are observed with a last maximum at a separation of ca $\lambda/4$. The signal decreases rather abruptly within the last 60 nm of the final approach. This abrupt decrease in scattering intensity is interpreted as the basic near field phenomenon. The range of this steep decay scales with the diameter of the aperture, reflecting the range of the near field of the antenna. The extent of quenching depends the polarisation of the exciting beam and is smaller for a purely dielectric object. Similar results are also obtained with pin antennas. In this case, however, conditions can be chosen such, that surface plasmons are excited in the pin antenna due to the choice of gold as a metal. This leads to a strong near field effect also for a dielectric object (14,26).

IMAGING BY SNOM

Fig. 3a shows a transmission electron micrograph of a 0.31 µm metal test pattern. Fig 3b shows an image of the 1 µm metal test pattern obtained with SNOM, using an aperture as an antenna. The triangular metal patches are clearly resolved at a resolution of about 50 nm corresponding approximately to the diameter of the apertures used. Fig 3c shows an image obtained with a pin antenna of a 0.31 µm S_iO_2 pattern. The optical signal was used to maintain a constant distance between the antenna and the object and the control signal was recorded as in the topographic mode of STM. In this case one can resolve the triangular patches but their form is not clearly revealed, although there are details in the image revealing a resolution of about 30 nm. We assume that the image reveals less clearly the evaporated pattern in this case, because the roughness of the underlying substrate also contributes to the recorded profile and prevents us from recognising the superimposed pattern.

Fig. 3a) b)

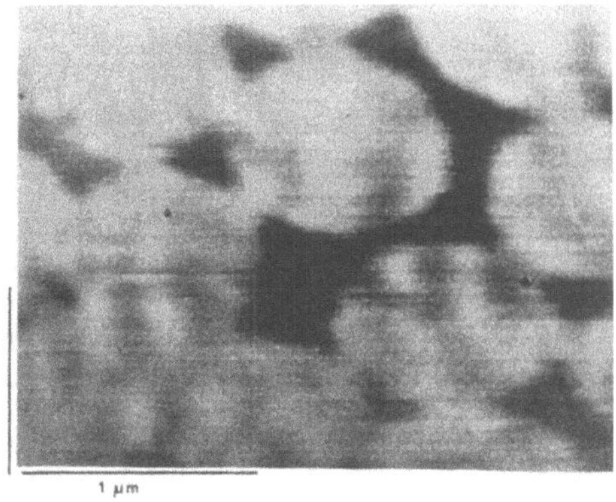

a) Transmission electron micrograph of a 0.31 µ metal pattern obtained by metal shadowing of a monolayer of 0.31 µm latex spheres and subsequent removal of the spheres (15).
b) SNOM image of a 1 µm metal pattern of 5 nm thickness using a 90 nm aperture. The optical signal was recorded and the distance between the object and the metal screen containing the aperture was kept constant using a tunnel current as a control signal.

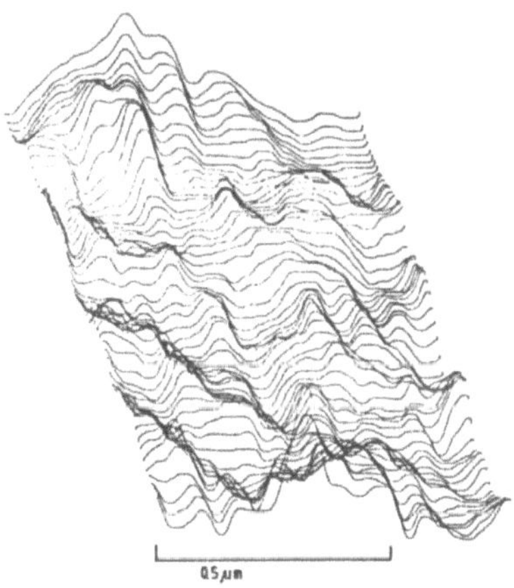

c) Image of a 0.31 µm S_iO_2 pattern obtained with a pin antenna (From (26)). In this case the optical signal is used to maintain a constant distance and the control signal is recorded similar in analogy to the topographic mode of STM.

OTHER TYPES OF SNOM

Near field phenomena can be clearly demonstrated in these reflection modes of SNOM. The location of the antenna in a planar metal film is however impractical concerning the accessibility of the antenna to the surface of a wider class of objects. Other realisations of SNOM, which are promising from this point of view all use as an antenna a specially formed tip at the end of a fibre serving as a waveguide. Pohl et al (5,16) used a quartz tip covered with an opaque metal film and a very small aperture at the apex of the tip as shown schematically in fig. 4a. The aperture is irradiated through the tip, and light transmitted through the aperture and a semitransparent object is

Fig 4

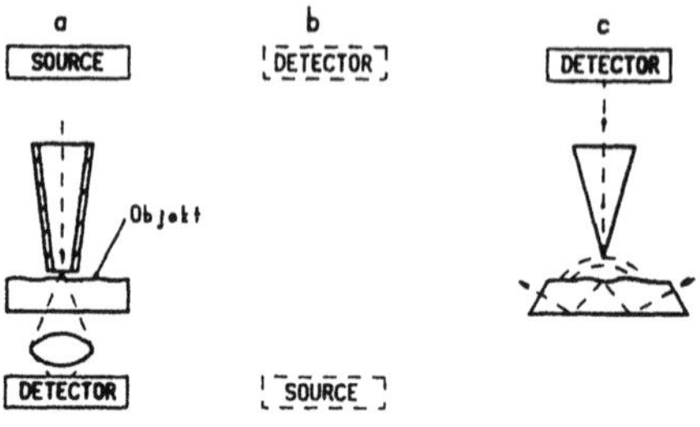

a) Transmission mode of SNOM, b) Collection mode c) collection mode with irradiation by evanescent waves.

detected by means of a light microscope and a photomultiplier, as the object is scanned laterally in close proximity to the surface of the object. With such a design optical images were obtained for the first time at a resolution in the range of 20 - 150 nm of thin film metal structures using visible light. An electronic tunnel current was used to maintain a constant distance during the scan (16).

Betzig et al realised an alternative transmission mode of SNOM, the collection mode (17), where the light path is reversed as compared to the one of Pohl et al as outlined in fig. 4b. They used an elegant and reliable method of probe fabrication. Sharply pointed pipettes are pulled from capillaries. They are then covered with an opaque metal film. In this way very smooth metallised tips with apertures of a diameter as small as 30 nm can be formed. The object is irradiated and light transmitted through the object is picked up by the aperture, fed into the capillary and recorded as the probe is raster scanned relatively to the sample at a distance of a few nm. Betzig et al performed their scans at a constant height without an additional control of the distance. As they perform subsequent scans at decreasing height, they observe an increase in lateral resolution of the obtained image, which is mainly limited by the size of the aperture. In addition, they demonstrated imaging of a fluorescent structure (18). An advantage of the collection mode is that no light microscope is needed in the detection pathway, facilitating a compact construction of the near field microscope.

A further variation of near field microscopy was introduced by Reddick et al (10) and by Courjon et al. (19). They modified the scheme of Betzig et al as shown in fig. 4c by changing the way of sample illumination and by using a simple dielectric tip at the end of a glas fibre as a probe instead of the metal covered tip with an aperture. The transparent sample is deposited on a flat glass substrate. The substrate is irradiated by total internal reflection such that evanescent waves are generated at its surface. The presence of the sample leads to a local variation of the evanescent light field as well as to a conversion of the evanescent waves into a radiation field. The tip is approached to the surface until it penetrates into the range of the evanescent field. The tip is excited by this evanescent field, and therefore some light is coupled into the fibre. For a nearly ideally flat specimen, the amount of light picked up by the fibre increases with the approach of the tip and the distance dependency of the light signal reflects mainly the penetration depth of the evanescent field and to some extent also on the form and size of the tip. In this way a purely optical signal of the distance between the tip and the substrate is obtained. This signal is used to control the distance between tip and sample during a scan and the control signal is recorded similar as in the topographic mode of scanning tunneling microscopy in order to obtain an image of the surface relief structure.

These methods which use a specially formed tip as a probe are rather limited in their application, because they dpend on a transparent object.

There seems to be a principal limitation of the imaging capability of SNOM in cases where object structures have dimensions in the order of the wavelength. Such objects may be considered as light guiding structures or as antennas and the imaging process will reflect the exitation profile of such structures rather than their surface topography. This effect, although of interest in itself, may distort images obtained by SNOM and limit its resolution. Such an effect would be comparable to influences of charge density waves on images obtained in scanning tunneling microscopy (20).

Apart from distorting effects of the object, other artefacts may be due to the design of the probe itself. In all cases of SNOM realised so far, contributions of radiation picked up from the near field range of the probe could not completely be separated from those picked up from background far field radiation, possibly due to an imperfect design of the probe (21).

We think that further substantial progress in the field of SNOM will depend on the proper design of tip shaped antenna suitable for SNOM in a reflection mode. The design of Reddick et al and Courjon et al was already modified by Spajer et al (22) to allow operation in a reflection mode. Instead of illumination of the object by total internal reflection, light is fed into the fibre and the light reflected from the tip is recorded. In this case a modulation of the distance between the dielectric tip and the sample allows to separate the modulated signal due to reflection from the tip from the DC background of other stray light.

SUGGESTIONS FOR NEW TYPES OF SENSORS

Due to the very nature of SNOM technique requiring to scan the probe in close proximity to the object, one has to be aware that occasional crashes between probe and sample or just a time dependent contamination of the probe may lead to a deterioration of its function . Therefore a simple, fast and inexpensive way of probe fabrication is required as it is in other scanning probe techniques. Part of the success of STM technology is due to the simple ways of forming sharp metal tips.

It is desirable to replace the dielectric tip used by Reddick et al by a tip being more efficient with respect to its function as a near field probe. We therefore require a concept for an efficient probe in conjunction with a simple technique for its fabrication.

We derive such a concept from a reconsideration of the basic aspects of SNOM and from an idea of probe fabrication dependent on the well established technique of drawing of glass coated metal wires (23)

The concepts of SNOM summarised above have in common, that the probe, be it an aperture or some kind or a specially designed tip, is connected to some kind of a cylindrical waveguide. Let us consider different kinds of cylindrical structures, capable of guiding electromagnetic waves:

Fig.5

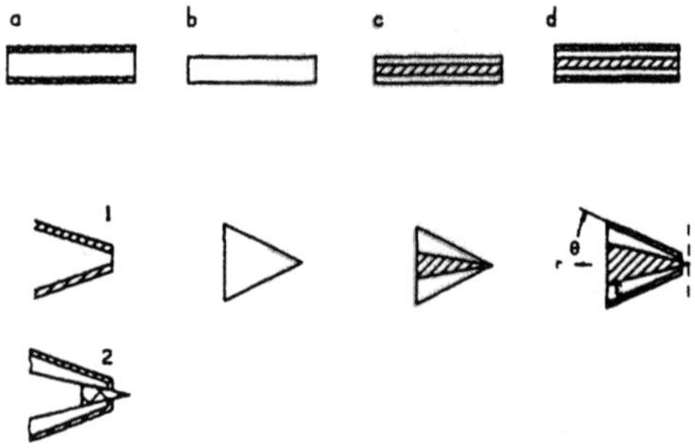

There is the hollow waveguide (fig. 5a) and a dielectric rod or fibre (fig 5b), both of which have been applied in SNOM as described above. The hollow waveguide has a cutoff frequency (24) and therefore is not expected to be an efficient device for coupling of electromagnetic energy into or out of a tip which is small compared to the wavelength. Lieberman et al (11) introduced a new approach to a probe for SNOM which tries to overcome this limitation. A small crystallite of an organic dye such as anthracene is grown into the open end of a pointed capillary as used by Betzig et al (Fig 5a2). By introducing this crystal the authors claim that electromagnetic energy can be channeled through a nano-aperture by exciton migration much more efficiently than by transmission through an aperture without the crystal.

A dielectric fibre has no cutoff frequency (24). Even if the diameter of the dielectric fibre is small compared to the wavelength, there exists a propagation mode. Therefore a dielectric fibre with a dielectric tip may seem to be a more reasonable choice for a probe for SNOM. The electromagnetic fields are however not tightly bound to the fibre but extend the more into the space outside, the smaller the diameter of the fibre is (24). It is therefore expected that no efficient localisation of electromagnetic fields to this kind of a structure is possible.

In this context attention should be drawn to other types of waveguides. A coaxial cable (fig 5d) is a well known wave guiding structure and the elctromagnetic fields are completely confined to the inner space of the coaxial line. A better confinement of electromagnetic energy as compared to a dielectric coating may also be obtainable by glass fibre with a metal core and without an additional metal coating (fig. 5c (24).

A conical coaxial line converging to a coaxial tip in theory supports a TEM mode (Eq. 1) which propagates out of or into the apex of the tip with a focusing effect (24):

$$(1) \quad E_\theta = (ik/2\pi fr\epsilon \sin(\theta)) \exp(+/-ikr)$$

E_θ is the amplitude of the electric field, f is the frequency. Polar coordinates are used (fig 5d). The values of θ extend only to the conical dielectric segment of dielectric constant ϵ. The focussing effect is due to the 1/r term in eq. 1.

These properties of a coaxial tip should make it ideally suited as a probe for SNOM. In addition it should be possible to fabricate this kind of a tip from "Taylor wires". Taylor wires are glass fibres with a metal core which are made by heating an appropriately chosen glass tube filled with metal to a temperature above the melting point of the metal and drawing of the capillary to a fine fibre (23). It should be possible to form tips from such fibres by local heating and pulling (25). We have started an activity to form such tips from fibres with a silver core. A metal coating of the tip could be done by metal deposition. It came to our attention that a coaxial tip as a sensor for SNOM was also suggested by Keilmann (29). He suggests however a rather different way of its realisation with a conical opening fabricated by micromechanical means into the solid flat end of a hollow waveguide, which would be difficult to implement in the optical domain.

The properties of the coaxial tip described above are valid only for the model of a infinitely conducting metal. At optical frequencies metals are far from fulfilling this condition. It remains to be shown how far the model is applicable. On the other hand surface plasmon modes exist in metals such as silver or gold at optical frequencies and it was shown that this property is very useful in the context of SNOM if a local surface plasmon can be excited in the antenna, especially if it is coupled to a delocalised surface plasmon such as a surface plasmon in a thin film (14,26). The concept of surface enhanced spectroscopy (27) can in this case be extended to near field microscopy. We expect that glass tips filled or coated with gold or silver will have interesting properties with respect to their use as a probe for SNOM, because it is well known, that surface plasmon modes exist in cylindrical geometries as well as spheres (28). In addition, these kind of probes could serve as a probe for SNOM and STM simultaneously.

APPLICATIONS OF SNOM

Once an efficient probe is found, SNOM may develop into an important tool for the microscopic characterisation of surfaces down to the 10 nm scale. On one hand it may be seen as an alternative to scanning force microscopy in the mapping of surface relief structures of non conducting surfaces. On the other hand SNOM is expected to have important unique applications in the material sciences, polymer - and photo - chemistry, as well in biology and technology:

1) Extension of microspectrophotometry of surfaces to the 10 nm scale. Localisation of chromophoric groups and dye lables in thin films, membrane fragments and fibre molecules.

2) Mapping of luminescent centers on surfaces.

3) Surface modification by local photochemistry.

4) Magnetooptics.

5) SNOM may turn out to be a unique tool to deposit information on - and read it out from - thin electrooptically active layers at a submicron scale. By applying a small voltage between the antenna and an electrically conducting surface as a support of the electrooptic layer, a strong electric field can be applied locally across the layer, leading to some electric field induced change in the layer. This change can be read out with the same antenna from the same spot.

This work was supported by the ESPRIT basic research action 3200 OLDS.

REFERENCES

1 O'Keefe JA (1956) J opt Soc Am 46:359
2 Bethe HA (1944) Phys Rev 66:163
3 Roberts A (1990) J Appl Opt 65:2896
4 Ash EK, Nichols N (1972) Nature 237:510
5 Pohl DW, Denk W, Lanz M (1984) Appl Phys Lett 44:651
6 Lewis A, Isaacson M, Harootunian A, Muray A (1984) Ultramicroscopy 13:227
7 Fischer UCh (1985) J Vac Sci Technol B3:386
8 Wessel J (1985) J Opt Soc Am B2:1583
9 Pohl DW, Fischer UCh, Dürig UT (1988) SPIE Vol 897:84
10Reddick RC, Warmack RJ, Ferrell TL(1989) Phys Rev B39:767
11Lieberman K, Harush S, Lewis A, Kopelman R (1990) Science 247:59
12Pohl DW (1990) In: Sheppard CJR, Mulvey T (eds) Advances in optical and electron microscopy. Academic Press, London
13Fischer UCh, Dürig U, Pohl DW (1988) Appl Phys Lett 52:249.
14Fischer UCh, Pohl DW (1989) Phys Rev Lett 62:458
15Fischer UCh, Zingsheim HP (1981) J Vac Sci Technol 19:881
16Dürig U, Pohl DW, Rohner F (1986) J Appl Phys 59:3318
17Betzig E, Isaacson M, Lewis A (1987) Appl Phys Lett 51:2088
20Heinzelmann H, Meyer E, Rudin H, Güntherodt HJ (1990)In: Behm RJ, Garcia N, Rohrer H (eds) Scanning tunneling microscopy and related methods. Kluwer Academic Publishers, Dordrecht Boston London
21Girard C, Spajer M (1990) Appl Optics in print
22Courjon D, Vigoureux JM, Spajer M, Sarayeddine K, Leblanc S (1990) Appl Optics in print
23Taylor GF (1924) Phys Rev 23:655
24Harriongton RF (1961) Time harmonic electromagnetic fields. McGraw-Hill, New York Toronto London
25Reitboeck HJ (1983) Journal of Neuroscience Methods 8:249
26Fischer UC (1990)In: Behm RJ, Garcia N, Rohrer H (eds) Scanning tunneling microscopy and related methods. Kluwer Academic Publishers, Dordrecht Boston London
27Metiu,H (1984) Progr Surf Sci 17:153
28Raether H (1988) Springer tracts in modrn physics 88
29Keilmann F (1988) German patent DE 3837389 C1

Study of epitaxial growth by combination of STM and LEED

M. Henzler, U. Köhler and O. Jusko

Institut für Festkörperphysik, Universität Hannover, Appelstraße 2, D-3000 Hannover 1

ABSTRACT

The first stages of epitaxial growth determine the growth of films with respect to general growth, orientation, smoothness, homogeneity and all other types of possible deviations from an ideal epitaxial film. For those studies it is important to see details of the starting substrate and of growing islands by direct microscopy, which is best accomplished by STM. The same way it is necessary to have good average values and distribution functions of sizes, distances and defect densities, which is best done by diffraction. It will be demonstrated, how the STM images from small areas and the diffraction patterns from large areas with low energy electrons (LEED) combine static informations on single defects and single islands with kinetic informations during growth at any temperature for a best description of homo-epitaxy and heteroepitaxy. The presence of defects like dislocations, stacking faults or domain boundaries is shown as their importance for nucleation. The kinetics of film growth is derived from diffraction measurements during deposition or annealing. Some defects like lattice constant variations (strain) or orientational defects (mosaics) are also seen in diffraction. In this way only a combination of techniques will enable a detailed description of growth processes.

1. Introduction

Crystal growth has been since long of general interest due to the regular plane faces of natural jewels. In modern technology artificial crystal growth is needed to provide single crystals. With the help of growing very thin layers on top of crystals with the same crystalline structure (epitaxy) many new materials, structures and consequently new semiconductor devices have been created. Since crystalline quality is decisive for the device quality, increased activities are in progress to study and understand the process of the formation of single atomic layers on top of a more or less perfect substrate. In an ideal description deposited atoms move freely until by chance a sufficiently high number coalesce to form a stable and growing nucleus. Smaller nuclei are likely to dissolve again. If a sufficient density of stable nuclei are formed, all incoming atoms are incorporated, until a complete layer is formed or new nucleation starts on top of the nuclei. In more real cases defects at the surface like steps or dislocations may act as preferential nucleation sites for one or many levels.

Therefore a lot of informations is needed for a fairly complete description of a growth process as simple and perfect as possible in an experiment. First of all a starting surface with the lowest number of defects has to be prepared. The starting surface has to be characterized with respect to all kinds of defects like contaminations or structural distortions. The arrangement of deposited atoms with respect to substrate defects and to each other has to be studied at any temperature and deposition rate to find preferential and stable configurations. Finally the completion of the first layer and the start of subsequent layers has to be observed. If the growing film has a composition different from the substrate (heteroepitaxy), even more growth features may happen due to different lattice constants and different binding energies. Since atoms move also after deposition, rearrangements and surface diffusion have to be studied. For a review see ref. 1.

Whereas technology usually considers only lengths in µm-scale or larger, solid state physics concentrates on atomic dimensions. Modern technology is down in nm-scale and modern solid state physics includes also lengths up to µm. So a good overlap is given. For surface studies a lot of techniques are available to cover the broad length scale. Some techniques are shown in Tab. 1 with the quite different ranges in vertical and lateral scale. The techniques may be divided in microscopy and diffraction techniques. The meaning of the abbreviations is as follows: SEM = Scanning Electron Microscopy, LEEM = Low Energy Electron Microscopy [2], TEM = Transmission Electron Microscopy, STM = Scanning Tunneling Microscopy, LEED = Low Energy Electron Diffraction, SPA-LEED = Spot Profile Analysis of LEED, RHEED = Reflection High Energy Diffraction [3]. It is the purpose of this paper to demonstrate the advantages, if techniques of the two groups are combined. In this case it will be Scanning Tunneling Microscopy (STM) and Low Energy Electron Diffraction (LEED).

2. Physical quantities to be measured

The above concept of growth studies on an atomic scale requires the measurement of a wide variety of physical quantities. Microscopy with an STM provides a lot of information, since both the starting surface and the growing and complete film are seen with all defects quite in detail. It is frequently, however, difficult to judge, if the observed area is typical for the total sample. For average values, size distributions and other quantitative values (like lattice constants and their variations) diffraction methods offer important additional and precise informations. Table 2 shows some advantages and disadvantages in a comparison. It should be pointed out, that recent developments of the STM with respect to speed and thermal stability (low drift) provide new possibi-

Table 1: Length scales for different techniques

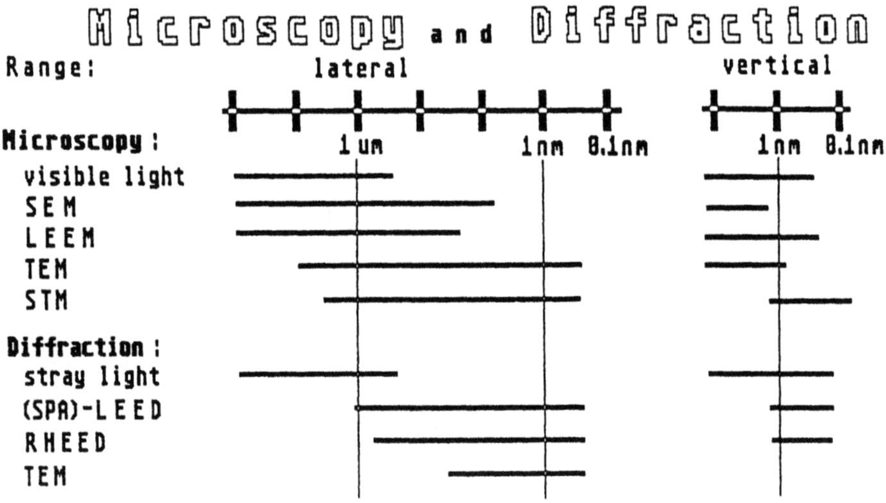

Table 2: Advantages and disadvantages for studies of different features

Comparison of STM and LEED

	STM	LEED
General features Existence of defects Identification of defects superstructure steps point defects islands domains	+	+
Qualitative features Single defect shape of island nucleation site correlation of defects rare events	+	−
Quantitative features Lattice constants and strain Average values Size distributions steps terraces islands domains	−	+
Kinetic features Changes during deposition Changes during heating annealing Reversible changes order disorder melting	−	+

lities for kinetic measurements in the near future. The combination of two techniques (one microscopy and one diffraction technique) is the best solution. Microscopy has another advantage, that images are in many cases easily interpreted, at least with respect to defects, islands and superstructure. Diffraction methods are more indirect. Instead of distances the pair correlation (the probability to find for a surface atom another surface atom in a given lateral and vertical distance) is all the information [4] (fig. 1). In a careful analysis the pair correlation provides a lot of informations. First of all the lattice constant including small variations by strain is precisely determined. All kind of lateral correlations like terrace width, island size and distance, domain size and so on are derived from a spot profile with average value and size distribution. Special distributions (like the geometric distribution) may point to the process of generation (like independence of width of neighboring terraces). An additional information is derived from the vertical pair correlation. Not only the step height (including multiple heights) is derived with high precision, the distribution of deposited atoms over different levels is derived from the energy dependance of the spot profile. If a different material is deposited, the coverage is also derived from the profile [5].

3. Examples for Homoepitaxy

The following examples will not be described in full details. Only those features are described, which show the special advantages of the techniques used to reveal some specialties of the growth process under investigation.

For the epitaxy of Si on Si(111) by molecular beam in uhv first the starting surface with the 7x7 superstructure is shown in fig. 2 [6]. It not only shows the wellknown superstructure, it also shows that as especially on semiconductors even the best surfaces always show an appreciable number of point defects. Those defects are also seen in diffraction as

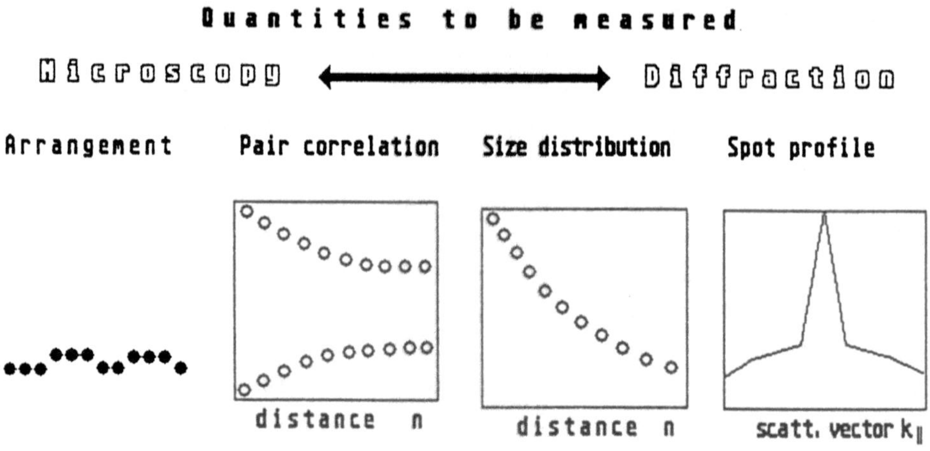

Fig. 1: Schematic presentations of the pair correlation as determined by microscopy and diffraction technique

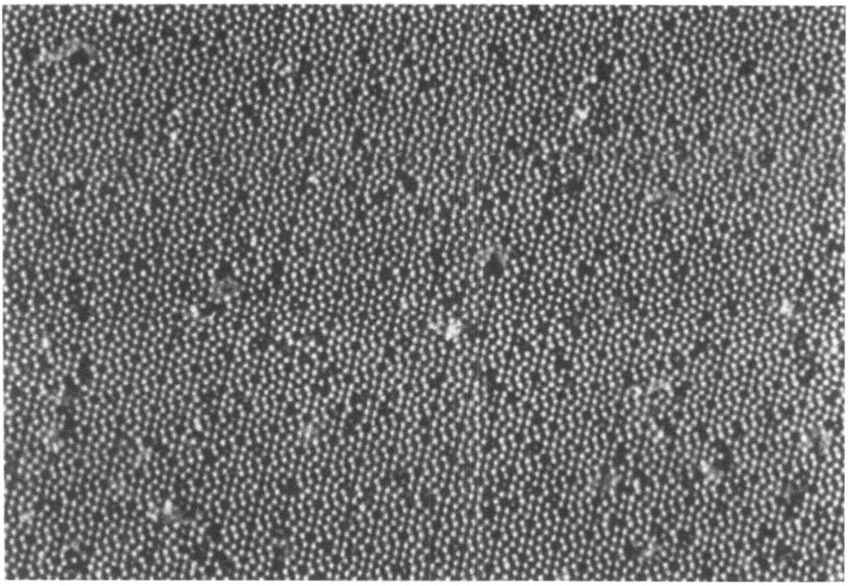

Fig. 2: STM image of a well annealed Si(111) with 7x7 superstructure. Even the best images show point defects [6].

background between the diffraction spots (fig. 3) [7]. In this case only the elastic background with the help of an electron monochromator and analyzer has been recorded. If onto a surface, optimized by high temperature annealing (1200°C), a fraction of a monolayer is deposited, islands are formed. Size and regularity increases with temperature. Fig. 4 shows an island after deposition at 550°C [6]. Shape and size distribution is seen in the spot profile due to the superposition of many islands. With diffraction the profile may be measured during deposition. Fig. 5 shows the oscillation of the peak intensity [8]. Each maximum corresponds to the completion of a layer. Whereas after two layers each layer is nearly completed, the first expected maximum at coverage $\theta = 1$ is missing. An analysis of the spot profile [9] reveals, that a fair fraction of the islands have double height, which is confirmed in fig. 3. The lower part of fig. 5 presents the increase of coverage in the levels 1, 2 and 3 during deposition. Here the fraction of islands with double height is easily seen [8]. Nucleation is more likely at defect sites. Fig. 6 shows the preferential nucleation at the domain boundaring of the substrate superstructure, which is clearly proven by the enlarged lower part [10]. Those correlations between defects and nucleation are only seen with microscopy.

For the Si(100) each plane has due to the diamond structure only twofold symmetry, which rotates by 90° from layer to layer. This symmetry is seen in fig. 7 by the direction of the dimer chains, which are rotated after crossing any step height of minimum height. It is also seen, that step edges parallel to chains are smooth, otherwise very rugged. If Si is deposited onto this surface, again dimer chains are formed (fig. 8). The high ratio of length to width decreases with increasing temperature [11]. In diffraction a superposition of the two orientations forms a fourfold asterisk, which is deconvoluted into the length and width distribution (fig. 9). More details on lateral and vertical distribution are found in ref. 12.

A last example for a surface without compositional variation is given with the roughening transition on the

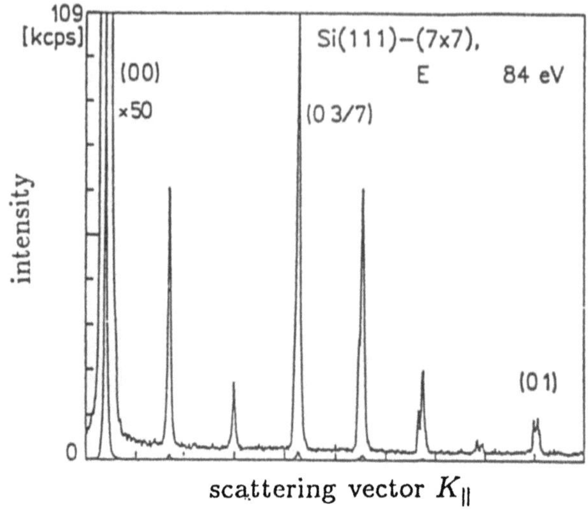

Fig. 3: Scan through normal and extra spots of the Si(111) 7x7. The background between spots is due to point defects [7].

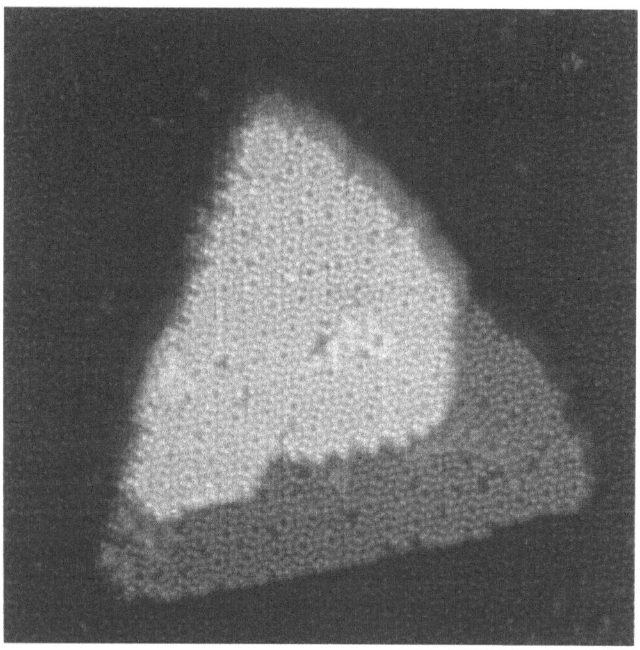

Fig. 4: Silicon island on a Si(111) after deposition of a quarter of a monolayer at T = 550^{0}C [6].

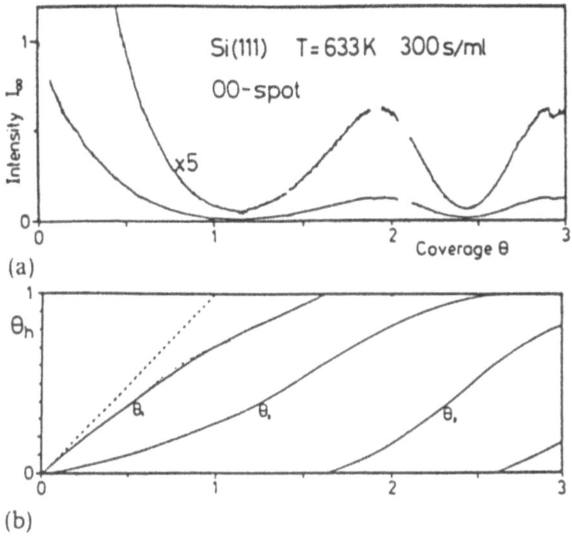

Fig. 5: Peak intensity (upper part) and layer distribution (lower part) during deposition of Si onto Si(111) [8].

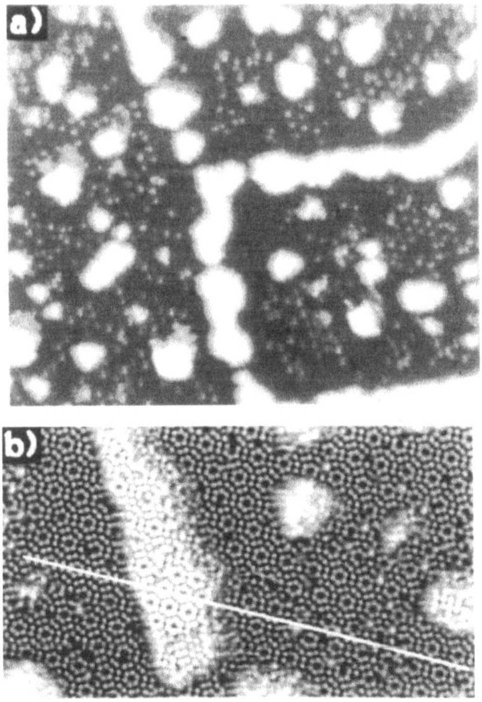

Fig. 6:

Nucleation at domain boundaries of the Si(111) 7x7 superstructure. The domain boundary under the island is seen from the enlarged lower section [10].

Fig. 7: STM image of a Si(100) with steps [11].

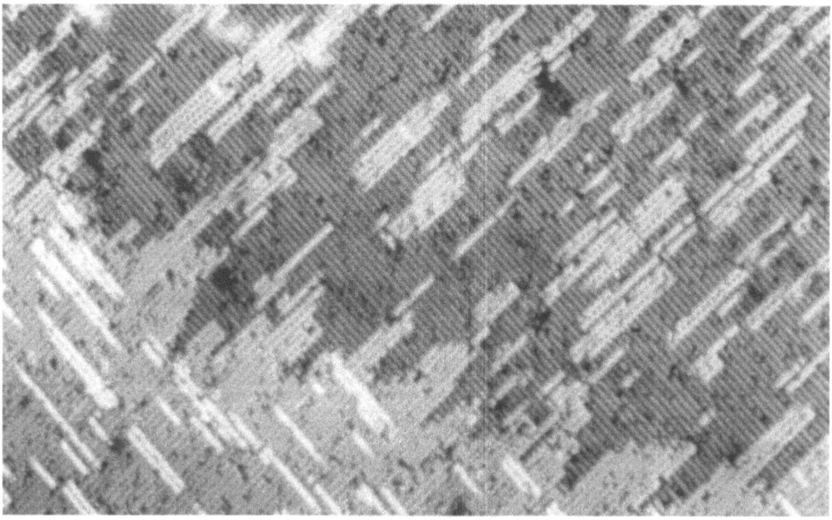

Fig. 8: Si(100) after depositon of a fraction of a monolayer of Si at about 100°C [11].

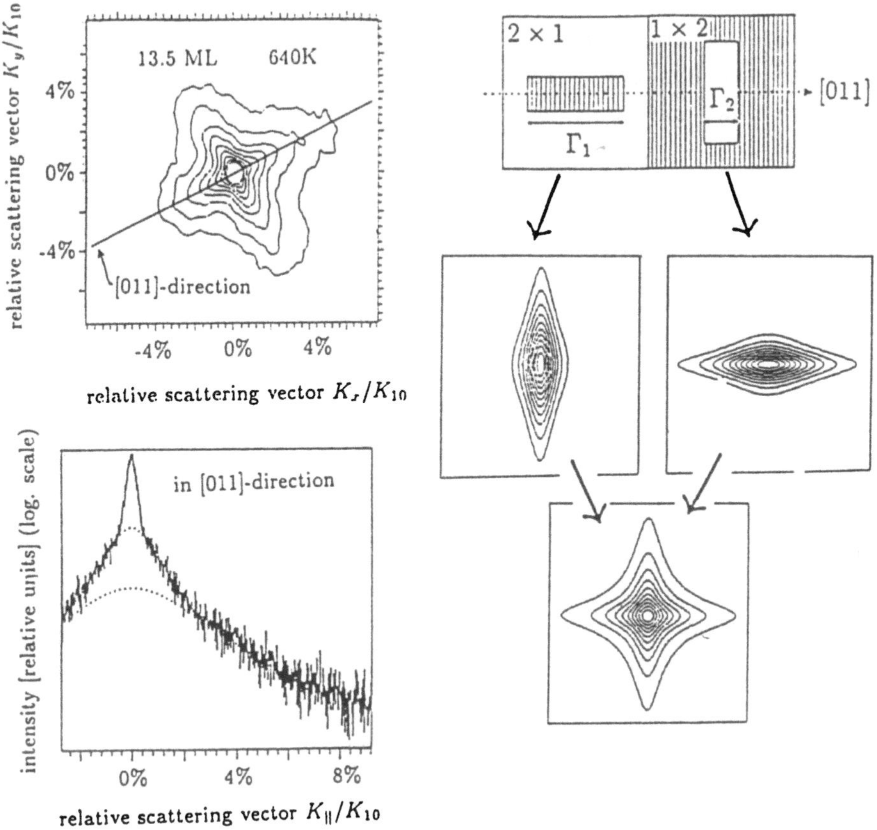

Fig. 9: Area scan (a) and cross section (b) of a LEED spot after deposition of 13.5 ML at T = 640 K, showing the asterisk shape and the composition out of central spike and two Lorentzian contributions. Part c shows the construction of the asterisk out of the two different types of islands.

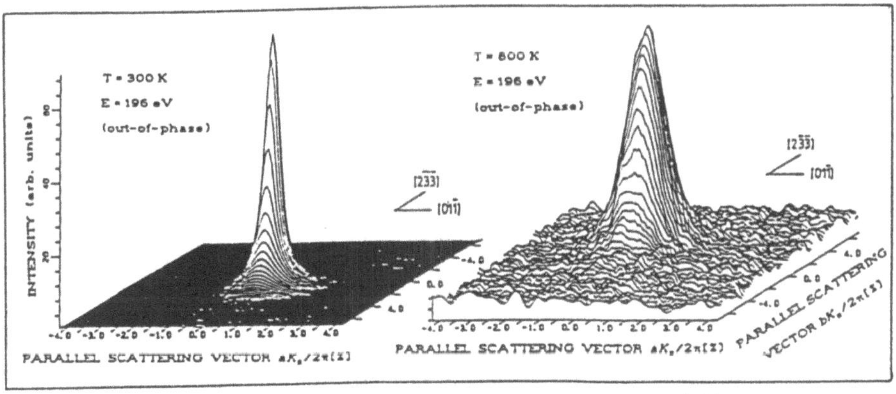

Fig. 10: Spot profiles of the Cu(113) at two different temperatures. The reversible change points to a roughening transition [13].

Cu(113). If the temperature is increased starting from room temperature, the diffraction spot profile increases in halfwidth reversibly (fig. 10). From energy dependence it is derived, that atoms move into lattice sites of higher levels, creating more steps and higher roughness. This reversible increase (with expected divergence at the roughening temperature) is shown in fig. 11. Those phase transitions so far are only observable with diffraction.

4. Examples for Heteroepitaxy

If a different material is deposited the new material may grow with same lattice constant and strain or with an independent lattice constant and a misfit. In the case of Pb on Cu(111) the Pb film up to monolayer grows with nearly the Pb bulk lattice constant and perfect orientation, as is shown in the diffraction spot (fig. 12). If it is heated close to 600 K (melting) a broadening in width and rotational angle demonstrates the melting by breaking up of the film into small domains with increasing rotational disorder [14]. If at room temperature additional Pb is deposited, the lattice constant of the film may be decreased by more than 3 % due to strain [14].

Whereas those thermodynamic and average properties are best determined by diffraction, special sites like single growth defects are best studied by microscopy. Some interesting features have been found with an STM study of the growth of Ge on Si(111) [10]. In fig. 13 a screw dislocation is seen (more precisely: a dislocation with a Burgers vector normal to the surface, since the direction of the dislocation may be inclined to the surface).

Screw dislocations are especially important for crystal growth, since no nucleation is needed here for continuation of growth. So far no STM images are available to demonstrate this process. Another defect is seen in fig. 14. Here two dislocations have opposite normal components of their Burgers vectors.

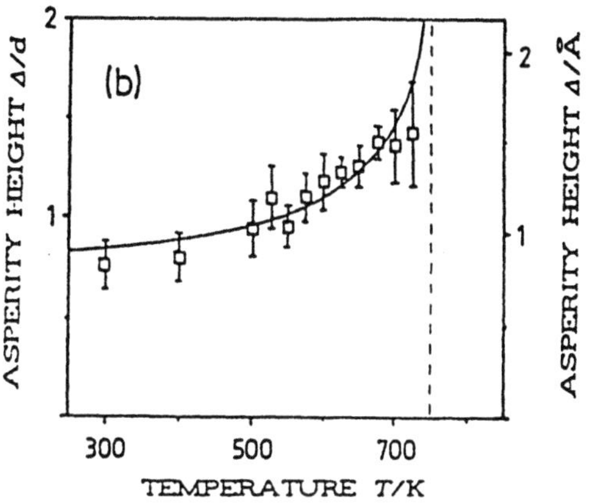

Fig. 11: Roughness analysis of the diffraction spots as shown in fig. 10 [13].

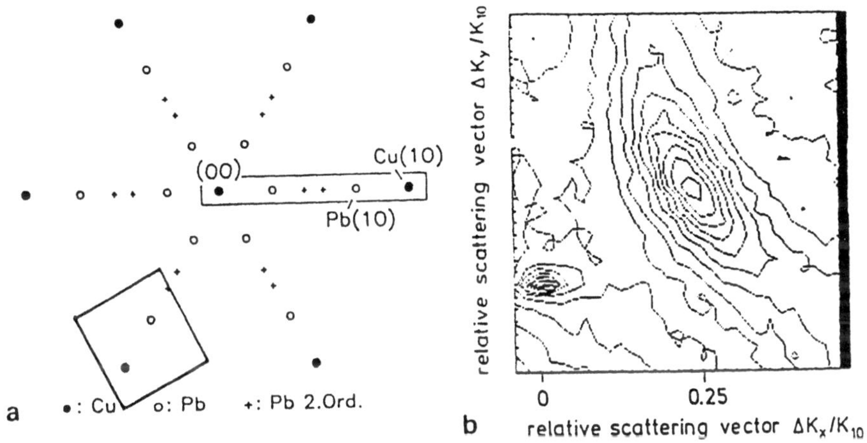

Fig. 12: a) Schematic diffraction pattern of a monolayer of Pb on Cu(111) at room temperature. b) Measured profile corresponding to the square of the lower left in part a) at T ≈ 600 K showing the melting of the Pb layer [14].

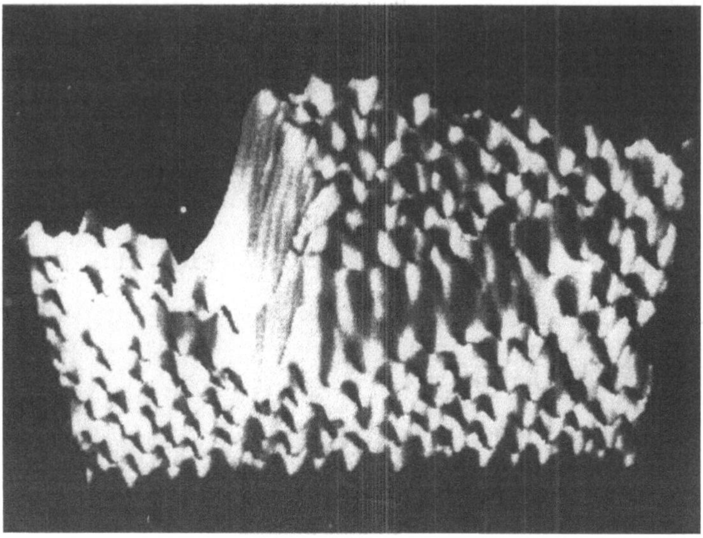

Fig. 13: Screw dislocation of a Ge film on Si(111) [10]. The vertical scale (step height 0.32 nm) is about 10 x the horizontal scale.

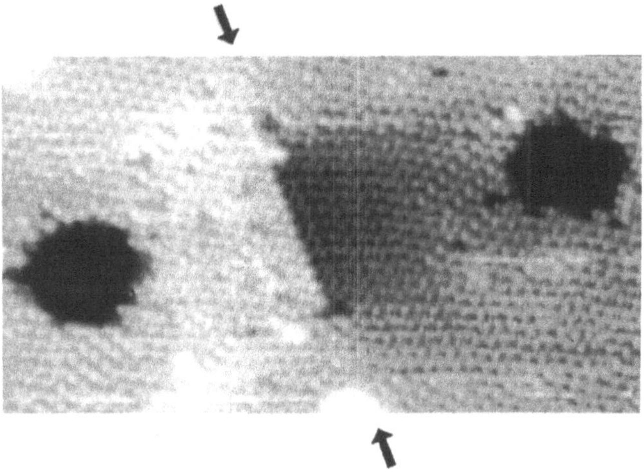

Fig. 14: Two partial dislocations of a Ge film on Si(111) at the end of the line between the arrows [10].

Fig. 15: Stacking fault (step height ≈ 0.1 nm) of a Ge film on Si(111) [10].

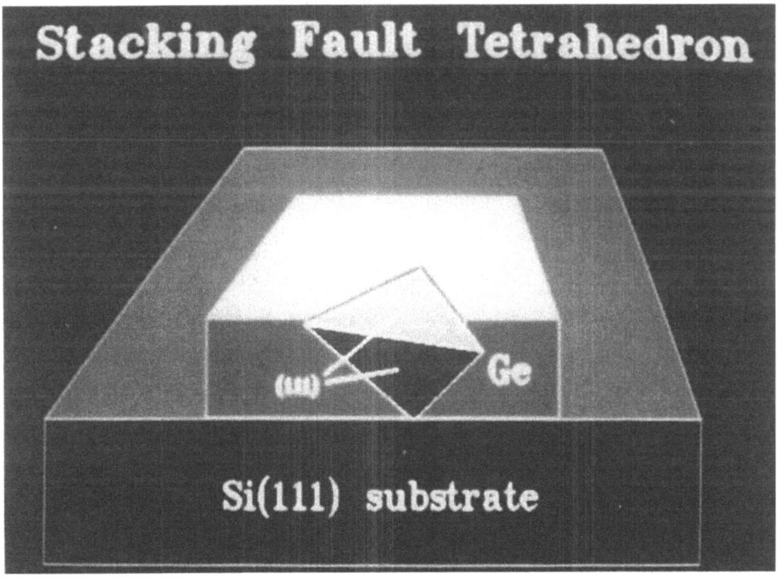

Fig. 16: Schematic presentation of the stacking fault in fig. 15.

More details are found as soon as the step height is measured. Since only about 0.1 nm is found, it cannot be an atomic step height with more than 0.3 nm height. Therefore the two disocations are partial dislocations which are generated by splitting a dislocation with a Burgers vector parallel to the surface. Therefore the plane between the partial dislocations is a stacking fault, so that the step height has to be 0.1 nm. Such a stacking fault is more clearly seen in fig. 15, which may form a complete triangle. Since the stacking fault follows always a (111) plane, the three stacking faults form a tetraeder starting at a defect at the interface (fig. 16). Therefore a careful study of surface defects reveals a lot of properties of the total film.

5. Conclusion

It has been shown, that both STM and LEED reveal a lot of important growth parameter, which are only in part available with either technique. A combination is therefore the best solution for a description as complete as possible.

Acknowledgements

The investigations have been supported by the Deutsche Forschungsgemeinschaft and the Stiftung Volkswagenwerk.

References

/1/ J.A. Venables, G.D.T. Spiller and M. Hanbücken (1984) Rep. Progr. Phys. 47: 399

/2/ E. Bauer, W. Telieps (1987) Scanning Microscopy Suppl. 1: 99

/3/ P.K. Larsen, P.J. Dobson, eds. (1988) Reflection High Energy Electron Diffraction and Reflection Electron Imaging of Surfaces, NATO ASI Series B, Vol. 188, Plenum Press New York

/4/ M. Henzler (1987) in: Festkörperprobleme Vol. 27, p 185, P. Grorse (ed.) Vieweg, Braunschweig

/5/ J. Wollschläger, J. Falta, M. Henzler (1990) Appl. Phys. A50: 57

/6/ U. Köhler, J.E. Demuth, R.J. Hamers (1989) J. Vac. Sci. Techn. A/7: 2860

/7/ H. Claus Diploma thesis, Hannover 1988

/8/ M. Horn von Hoegen, J. Falta, M. Henzler (1989) Thin Solid Films 183: 213

/9/ R. Altsinger, H. Busch, M. Horn, M. Henzler (1988) Surf. Sci. 200: 235

/10/ U. Köhler, O. Jusko, to be published

/11/ J.Y. Tsao, E. Chason, U. Köhler, R. Hamers (1989) Phys. Rev. B40: 11951

/12/ S. Heun, J. Falta, M. Henzler (1991) Surf. Sci. in press

/13/ J. Wollschläger, PhD thesis Hannover 1990

/14/ G. Meyer, M. Michailov, M. Henzler (1988) Surf. Sci. 202: 125

STM STUDIES OF ADSORBATES IN THE MONOLAYER RANGE: Ag/Ni(100) AND O/Ni(100)

A. Brodde, G. Wilhelmi and H. Neddermeyer

Institut für Experimentalphysik der Ruhr-Universität Bochum,
Postfach 10 21 48, D-4630 Bochum, Fed. Rep. Germany

ABSTRACT. We have measured Ni(100) in its clean state and the systems Ag/Ni(100) and O/Ni(100) in the monolayer coverage range by using STM, LEED and AES. The STM images from clean Ni(100) very often show a very complicated terrace structure due to in general monoatomic steps, which is related to the misorientation of the surface against (100) and to bulk imperfections. For Ni(100) an atomic corrugation of less than 0.1 Å was found. For Ag deposition at room temperature we observe in the submonolayer coverage range a network of connected islands of monoatomic height, whose shape can surprisingly well be explained by diffusion-limited aggregation. The step edges are the preferential sites for formation of the Ag nuclei. Two domains are identified on the Ag islands which are consistent with the c(2x8) reconstruction visible in LEED. For the system O/Ni(100) we have identified p(2x2) and c(2x2) local structures depending on the preparation conditions and O exposure. At room temperature we have also found atomic features, which we assign to NiO nuclei. A kind of epitaxial NiO layer has been obtained by O exposures of 5000 L at a substrate temperature of 270 °C. In STM the NiO layer appears as a fairly rough surface structure with a typical corrugation of 50 Å. It consists of smaller NiO islands, which very often show an ordered arrangement along high-symmetry directions but do not exhibit a distinct atomic structure.

1. INTRODUCTION

The use of scanning tunneling microscopy (STM) for the study of initial stages of film growth [1] and of adsorption on metal [2] and semiconductor substrates [3] belongs to the most promising fields in this new research area. The great advantage of STM for such work is its access to random effects on an atomic scale. Information on the initial adsorption sites and on the growth of the adlayer structures, where a long-range order has not yet been developed, may directly be determined from the STM images. It has to be emphasized, however, that such STM experiments turn out to be rather difficult in some cases. For a reactive metal as Ni(100) the first difficulty is the preparation of a satisfactory and clean substrate surface. While the low-energy electron diffraction (LEED) pattern still seems to indicate the presence of a well-ordered surface, it might not even be possible to obtain a well resolved STM pattern due to an apparent roughness of the surface, which might be caused by extensive bombarding and annealing cycles. A second difficulty is the necessary stability of the STM and the particular tip in use to obtain atomically resolved images from metal surfaces. Since the atomic corrugation of a metal surface is normally (i.e., without enhancement effect [4]) in the order of 0.1 Å (or smaller) it is only resolved under optimum conditions of the instrumental equipment.

The interpretation of the results is not straightforward, since the local electronic structure and the adsorption geometry of the adsorbed species are not known, a priori, and both contribute to the STM images. Moreover, the observed effects are usually more complicated than expected on the basis of the LEED pattern. The exposure to reactive molecules may give rise to various features in the images due to simultaneous formation of different surface compounds.

The present work is devoted to Ni(100) as substrate and to the adsorption of Ag and O. Ag/Ni and O/Ni are typical representatives for a non-reactive and a reactive adsorbate/substrate system. A comparison between such systems is particularly useful for the evaluation of electronic effects in the STM images. In the case of a strong adsorbate/substrate interaction the local electronic structure might be changed considerably in comparison with the clean substrate surface or with a system, where the charge transfer between the components is small. These differences may then lead to the appearance of a protrusion on the site of the adsorbed species in the latter case, while for a reactive system like O/Ni the chemisorbed O atoms are connected with a depression in the image.

The atomic radius of Ag (1.45 Å) differs by 16 % from that of Ni (1.25 Å) [5] and interdiffusion is unlikely. Due to the large lattice mismatch of Ag and Ni the Ag islands do initially (i.e., in the first atomic layer) not grow in form of Ag(100) but rather in a slightly compressed hexagonal close-packed structure [6,7]. In LEED a c(2x8) pattern is obtained at this stage of coverage [7]. The formation of a hexagonal surface layer on top of a square lattice is a phenomenon frequently found for metal systems [8] and qualitatively indicates that the interaction between the adsorbed Ag atoms is stronger than between the Ag and Ni atoms. STM measurements of Ag/Ni(100) have not yet been reported in the literature, to our knowledge.

The recent interest in the system O on Ni(100) concentrates on the low-coverage regime, where a fully ordered adsorbate/substrate structure has not yet been formed [9-11] and on surfaces, where a thin layer of NiO has been grown [12-15] by using O exposures in the order of 1000 Langmuir (L) at temperatures of a few 100 °C (1 L is approximately equivalent to an exposure at 10^{-6} mbar for 1 s). The low-coverage studies of Ref. 9

and 10 have been performed at low temperatures (80 K) and the diffraction pattern did not indicate the presence of long-range order. In the diffuse low-energy electron diffraction (DLEED) study of Ref. 9 the adsorption sites for O were found slightly shifted by 0.4 Å against the ideal 4- fold hollow sites of Ni(100), which could eventually be visualized by STM. In Ref. 11 an unexpected adsorption geometry was found by Monte Carlo simulation, which should give rise to a c(2x2) diffraction pattern although the local geometry was of the p(2x2)-type. Previous STM measurements on O/Ni(100 have been performed by Binnig et al. [16] and Behm [17].

In the present paper some experimental details on the instrumental equipment and on the sample preparation will be given in the following chapter. The main part of this paper will be a description of the results from clean Ni(100), the systems Ag/Ni(100), O/Ni(100) and a discussion in the light of other surface physical measurements of these systems.

2. EXPERIMENT

The instrumental equipment used for the present measurements consists of a main UHV chamber containing STM, LEED system, electron energy analyzer for AES, ion gun for sputtering, heating stage for annealing the samples, effusion cell for Ag evaporation and gas inlet system for oxygen and rare gases. A transfer rod to a vacuum lock allows quick sample introduction and exchange of the tunneling tips. The base pressure of the system is in the 10^{-11} mbar range. More details of the apparatus have been described elsewhere [18].

The Ni(100) substrates were oriented within an accuracy of 0.5 °...1 °. After mechanical and chemical polishing the Ni(100) substrates were cleaned in situ by repeated cycles of argon bombardment and annealing up to 800° C until in the Auger electron spectra contamination by C, S and O could no longer be detected within the noise of the data. A clear LEED pattern with 4-fold symmetry according to clean Ni(100) was finally observed. The adsorption experiments were performed in the STM chamber.

For the STM measurements constant current topographies (CCT's) and local I/U characteristics have been recorded for sample bias voltages between -1 and 1 V and a tunneling current in the range of 1...20 nA. For the clean surface and in the low-coverage range (up to an Ag coverage of 1/3 monolayer (ML) and an O_2 exposure of 50 L) atomically resolved images could be obtained, although a large number of experiments have been found necessary to obtain satisfactory results. In many of the O_2 experiments, however, the presence of oxygen on the sample surface and on the tip led to instabilities in the tunneling current or to insufficient sharpness of the tip and to images without atomic resolution. For clean Ni(100) and under optimum conditions an atomic corrugation due to the Ni atoms of less than 0.1 Å was observed. The rows of atoms appeared as straight lines and the entire surface showed the presence of defect-like structures in form of depressions with a concentration of about 1 % of a monolayer. The position of these defects appeared in 4-fold hollow or in bridge sites and can therefore not be caused by missing Ni atoms. Instead, we assign them to locally reacted Ni atoms due to contamination by C, O or S.

Ag was deposited onto the room temperature Ni substrates by using a water- cooled Knudsen-type effusion cell with a rate of 1/5 ML/min and a pressure of approximately 3×10^{-10} mbar. The monolayer scale is as usual referred to the number of Ni substrate atoms in the first atomic layer (i.e., $\Theta = 1$ is equivalent to a density of 1.61×10^{15} atoms/cm^2. Since the density of Ag atoms in a dense-packed Ag(111) layer is only 1.38×10^{15} atoms/cm^2 for $\Theta = 1$ the second atomic layer of Ag (assuming the bulk fcc (111) geometry) should already cover 17 % of the surface. The absolute accuracy of the ML scale as monitored by a quartz oscillator is estimated to be 5...10 %. For $\Theta = 1$ we observed two-dimensional Ag islands in the second layer position covering 24 % of the surface in good agreement with the nominal coverage.

The macroscopic topography of the surfaces, i.e., the terrace and step structure could always be resolved. Because of the limitations in the orientation for monoatomic steps a mean terrace width of 200 Å was expected and actually observed. On some parts of the surface additional steps were found due to lattice imperfections in the bulk. These steps cross the misorientation-related steps mostly without any noticeable interaction and appear normally completely straight in a low-index symmetry direction of the crystal. In some cases a screw dislocation was found to be the origin of these steps. Since steps provide preferential sites for Ag condensation and oxide growth (see below) the possible existence of a rather complicated step structure on Ni(100) has to be considered for interpretation of the results.

3. RESULTS AND DISCUSSION

We start with a description of the results obtained on clean Ni(100) and continue with the systems Ag/Ni(100) and O/Ni(100). The topographical STM measurements (CCT's) are shown as gray tone images. Bright areas correspond to protrusions and dark areas to depressions as seen by the tunneling tip. An oblique plane has always been subtracted from the raw data to utilize the entire range of gray tones for display of the individual atomic structures or the arrangements of terraces. To visualize atomic structures on different terraces, which in this case are mostly formed by a monoatomic step structure, a contrast enhancement technique has been applied to the results similar to that described in Ref. [19]. Otherwise the small atomic corrugation (at most a few 0.1 Å) cannot be represented with sufficient contrast in comparison with that caused by the steps. Fourier filtering has been applied to one CCT from clean Ni(100) (see below), where the noise in the raw data (a few 0.01 Å) came close to the atomic corrugation (0.05...0.1 Å).

Ni(100)

In Fig. 1 a CCT from clean Ni(100) is reproduced, which represents one of the typical terrace structures obtained on this surface. Each line separating darker and brighter areas correspond to a monoatomic step. Some of the edges marked with a "v" run remarkably straight, while others (marked with a "w") show distinct deviations from straight lines. Obviously, the lines form arrays of the same (or in case of the w-steps similar) orientation and mostly intersect without noticeable interaction. In some images (not shown here) v-steps are terminated at w-steps or at screw dislocations. We explain this complicated behavior by a combined effect of the macroscopic misorientation of the sample and by the presence of glide planes in the bulk, which are generated by strain fields in the sample. In fact we observe the strain-related step structure even be different in subsequent measurements of about the same surface area and in other cases completely absent.

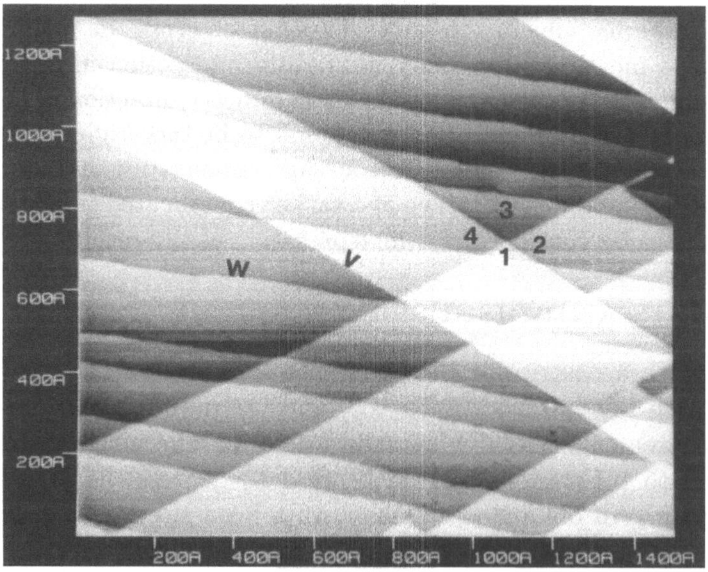

Fig. 1. Overview image from Ni(100) obtained at a sample bias voltage of $U = 50$ mV (sample positive) and a tunneling current $I = 10$ nA.

Annealing of Ni(100) may lead to drastic changes of the topography near a corner of intersecting steps (e.g., at corners like that defined by terraces 1, 2, 3 and 4 in the image), where the upper (1) and lower (3) terrace withdraw from each other by growing of a corridor-like connection of terraces 2 and 4 (see below results from O/Ni(100)). The different gray tone levels provide the impression of "transparent" surface areas but are as a matter of fact a consequence of the independent origin of the observed step arrays.

Fig. 2. Atomic corrugation on Ni(100). Fourier filtering has been applied to the raw data. The area is $\approx 23 \times 23$ Å^2.

The atomic corrugation of Ni(100) could only be identified under the experimental conditions of a specifically small noise in the $z-$value of the tunneling tip. It also turned out that the square lattice of Ni surface atoms normally showed a large number of defect structures (appearing as atomic depression) in 4-fold hollow or in bridge sites. Since their mean concentration over a large surface area corresponded to approximately 3 % of one ML (see Ref. [15]), we assign these defects to surface contamination by O, C or S atoms. These defects are even visible in the results shown in Fig. 1, if the gray tone scale is optimized. Since the noise very often appears as a streaking in the scanning direction (along the $x-$axis of the images) we have applied Fourier filtering to one CCT and show the result in Fig. 2. The mean corrugation in this image is 0.1 Å. The atomic features clearly show the expected periodicity of a square lattice. By comparison with atomically resolved images containing atomic defects and using the known orientation of the sample we are able to deduce, where the atomic nuclei of Ni atoms are located laterally. In this case they are located below the protrusions as one would intuitively expect on the basis of simple charge density arguments. We should mention that an enhancement mechanism for the observed corrugation as was found, e.g., for Al(111) [4] is probably absent in the case of STM measurements on Ni(100).

Ag/Ni(100)

We first describe results obtained in the low-coverage range. In Fig. 3 an image is reproduced from a surface, where 1/6 ML of Ag has been deposited on Ni(100) being at room temperature. Several terraces are visible, which are separated by monoatomic steps. In this case the step structure is different than that shown in Fig. 1 and is entirely due to the local deviations from ideal Ni(100) by macroscopic misorientation or the polishing procedure of the substrate. The Ag deposit is easily recognized in form of two-dimensional islands with fairly irregular shape. The measured height above the Ni substrate is 2.6 Å. The island density is particularly large at steps, which are sometimes decorated completely with the Ag deposit. Since monoatomic Ni steps between neighboring (100) terraces have a height of 1.75 Å (we measure a height of 1.8 Å), the Ag islands are higher than an uncovered (upper) Ni terrace due to the larger size of the Ag atoms. This also indicates that electronic effects upon Ag adsorption can only be of minor importance.

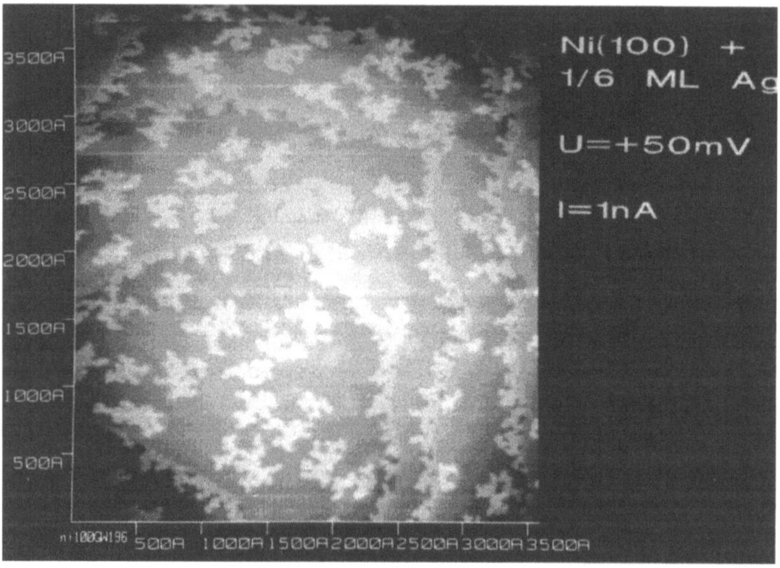

Fig. 3. Overview image from 1/6 ML Ag on Ni(100).

The general shape of the Ag islands is characterized by their open nature. In contrast to Ag islands on Si(111)7x7 [20] they do not show a tendency of minimizing their edge length but rather form isolated networks of smaller two-dimensional Ag subunits with typical lateral extension of 50 Å. The growth of these islands upon further Ag deposition continues by increase of their lateral extension without loosing their characteristic structure (not shown here). The shape of these Ag islands is in striking agreement with that found in diffusion-limited aggregation (DLA) of particles [21]. In DLA an initial "seed" particle is present on some position of the substrate. A second particle is allowed to walk at random from far away until it arrives at one of the sites adjacent to the "seed" particle and stops. Another particle is then diffusing on the surface and halted adjacent to the two interconnected particles. This process may continue until single particles are no longer formed. An interesting consequence of the good agreement between DLA growth and the present results would be that the diffusing particles are not Ag atoms but larger subunits, which, however, seems to be difficult to understand. We therefore suggest that the growth of a hexagonal Ag overlayer structure on a square lattice may also contribute to the observed general shape of the Ag islands. Since with increasing size of Ag subunits strain fields within the Ag layer also increase, it seems plausible that the observed typical width of 50 Å of the subunits corresponds to a balance of energy gain by adding further Ag atoms to existing subunits with the increase in strain energy.

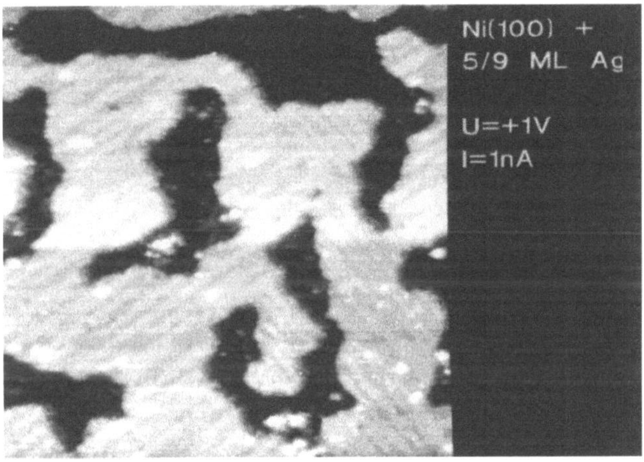

Fig. 4. Domain structure on Ag/Ni(100). The area is ≈ 450 × 450 $Å^2$.

Microscopic details of the growth morphology are resolved in the CCT reproduced in Fig. 4. On top of the monoatomic Ag islands a weak pattern of lines (occurring in orthogonal domains) is identified. These lines are oriented in [110]-like directions and therefore follow the directions of the c(2x8) unit cells [7]. The LEED pattern from our surface is in agreement with the results reported in Ref. [7] and it is obvious that the arrangement of lines reflects the modulation of the surface due to growth of a nearly hexagonal Ag overlayer on Ni(100).

For $\Theta = 1$ the first atomic Ag layer should be completed and part of the second layer already be grown due to the differences of the atomic radii of

Ag and Ni. This is confirmed by the experiment (Fig. 5). The terraces are now completely covered with Ag and the surplus Ag atoms are condensed in form of two-dimensional islands in second layer position. Note that their form may principally also be explained by DLA although their general shape seems to be quite different. Actually, the main difference to the submonolayer coverage range is that the mean lateral extension of the two- dimensional Ag aggregates now has increased to nearly 200 Å. The connection of such Ag aggregates to larger Ag islands is clearly visible in the CCT shown in Fig. 5.

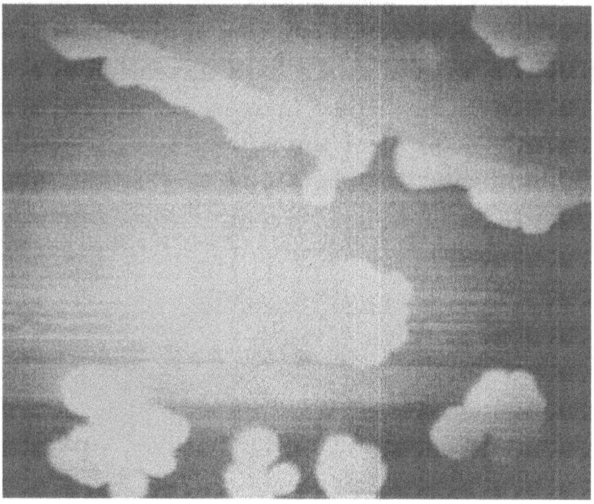

Fig. 5. STM image from 1 ML Ag on Ni(100) obtained at $U = 50$ mV and $I = 1$ nA. The base is covered completely with Ag, the brighter parts are two-dimensional islands of Ag, which due to the different size with regard to the substrate lattice could not be incorporated in the first atomic layer. The area is $\approx 1700 \times 2100$ Å^2.

O/Ni(100)

The ideal p(2x2) O/Ni(100) structure corresponds to an O coverage $\Theta = 0.25$, whereas c(2x2) may accommodate twice as many O atoms [9-11]. Consequently, with increasing O exposure in the initial stage of ordering a p(2x2) and later a c(2x2) structure is expected in LEED and STM. This general behavior was confirmed by the present experiment if the substrate temperature was high enough to lead to long-range order in the adsorbate. We did not investigate the temperature dependence of the formation of ordered adsorbate structures more systematically but generally found that in STM at room temperature and up to an exposure of about 10 L O tends to form local p(2x2) unit cells and other O-induced random surface structures, which we assign to NiO nuclei. On the other hand, for a hot substrate (typically a few 100 °C) a long-range order p(2x2) structure is only found in STM for small exposure (typically 1 L) which after further exposure is followed by formation of c(2x2) unit cells. Other random O-induced features are hardly visible in low-coverage experiments above room temperature (see below).

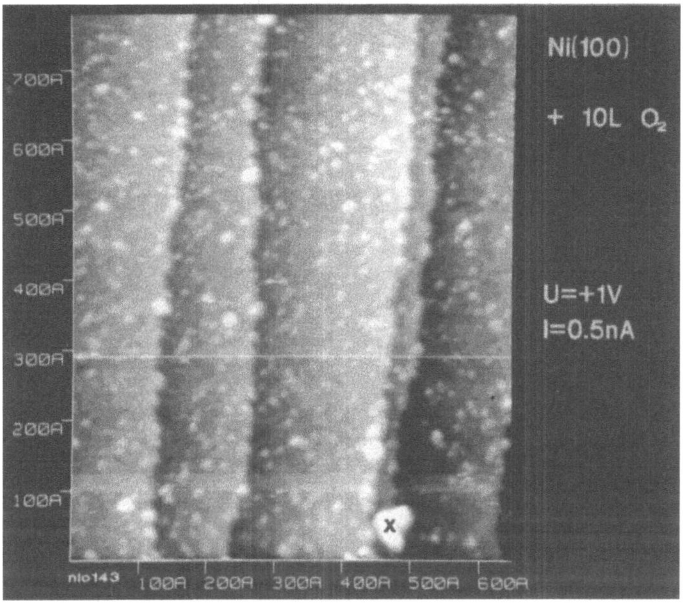

Fig. 6. CCT from 10 L O_2 on Ni(100) exposed at room temperature.

We first describe results from room temperature adsorption. In Fig. 6 a CCT has been reproduced, which demonstrates the occurrence of gross effects of O exposure. The image shows a series of monoatomic steps due to the misorientation of the sample. The influence of O adsorption on the surface leads to the appearance of protrusions on the entire surface, which do not exhibit long-range order. The size of these protrusions may differ considerably, the largest have already grown to islands with lateral dimensions in the order of 50 Å (e.g., at x) and we ascribe these features to NiO islands. The NiO islands grow preferentially near step edges. The LEED pattern of this surface showed a clear c(2x2) symmetry, extra p(2x2) beams were hardly visible. In this particular case we did not find ordered surface structures with STM probably due to noise and lack of sufficient resolution.

However, on a smaller surface area and after the tip has been conditioned for atomic resolution, images like that shown in Fig. 7 could be obtained. The image contains two monoatomic steps (marked with 1 and 2). Although the arrangement of atomic species still does not exhibit long-range order, regular building blocks in a sometimes periodic arrangement are clearly visible. The local structure is generally of the p(2x2)-type. The more protruding features in the image correspond to NiO nuclei, to our opinion. Note that the above mentioned contrast enhancement technique has been applied to the image in order to visualize the atomic effects of the different terraces more clearly.

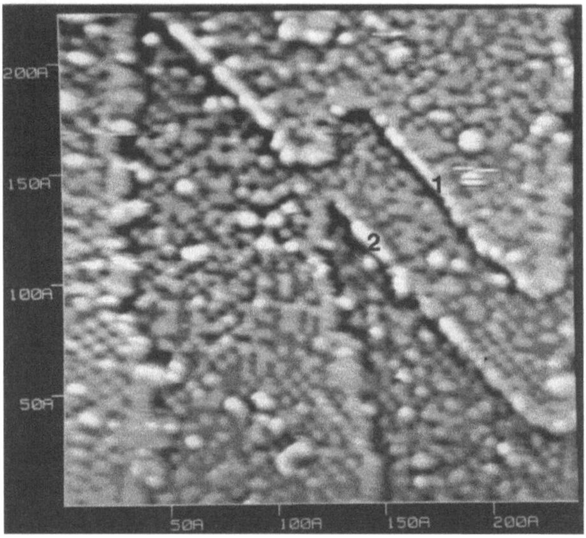

Fig. 7. Atomically resolved CCT from 10 L O_2 on Ni(100) exposed at room temperature. $U = -1$ V and $I = 2$ nA.

The LEED pattern of the sample used for the STM measurements shown in Fig. 7 was of c(2x2) symmetry. This could be explained in terms of a low- coverage c(2x2) phase, which results from two domains of local p(2x2) structures shifted against each other by the Ni lattice parameter. The Fourier transform of a similar STM image indeed showed a clear c(2x2) symmetry. Qualitatively, these results are in agreement with the prediction of Bartelt et al. of an unexpected low-coverage c(2x2) phase [11] . However, in contrast to the model of Ref. [11] in Fig. 7 the formation of c(2x2) is already visible locally, i.e., ordered species are already found in a distance of $\sqrt{2}$ times the Ni lattice parameter.

In Fig. 8 an example for the presence of c(2x2) unit cells is shown. The differences to the p(2x2) symmetry (Fig. 7) are recognized by the different orientation of the unit cells (they are rotated by 45° against those seen in Fig. 7) and by the smaller lattice parameter. The general arrangement of the c(2x2) unit cells shows a better order than that of the p(2x2) unit cells but seems to be far away from one expected on a surface with long-range order. Nevertheless, this sample produced a satisfactory c(2x2) LEED pattern. Note that the additional protrusions due to NiO nuclei are not present on the surface. This observation correlates with the fact that the more dense c(2x2) structure is obtained on this surface and not the p(2x2) structure, although the O exposure was smaller than in case of the sample used for the measurement shown in Fig. 7.

Fig. 8. Atomically resolved CCT from O/Ni(100) showing c(2x2) unit cells.

An important point in the low-coverage regime is that of the localization of the O atoms. A number of reasons may be given for an assignment of the adsorbed O atoms to the depressions visible in Figs. 7 and 8. Firstly, their density increases with increasing O exposure. Secondly, similar depressions are measured on "clean" Ni(100) on 4-fold hollow sites. These depressions are very likely caused by residual contamination of the surface (their concentration is consistent with the possible contamination by O, S and C). In a naive picture of the electronic effects upon O adsorption Ni states near the Fermi level E_F are depleted and transferred to O. The reduction of states near E_F is then responsible for the occurrence of a depression in the image. We note that it is also known from other surfaces exposed to O that the initial sites for O appear as depressions and not as protruding features [17].

We finally describe an STM image from a surface which has been exposed to 5000 L of O at a temperature of 270 °C (Fig. 9). According to the literature a thin Ni oxide layer with (100) orientation should then be grown on the surface. The LEED pattern measured on this surface still showed the integral order beams of Ni(100) and those of c(2x2) symmetry. The presence of epitaxial NiO was indicated by the expected additional beams, which, however, showed considerable broadening. The STM image (Fig. 9) is characterized by an appreciable roughness, which means that in fact not a thin oxide layer has been grown under these conditions but rather a large number of three-dimensional NiO islands. Some of the islands show registration with main symmetry directions of the substrate (see position marked with "x"), which indicates regular growth in the initial stage. Due to the large lattice mismatch between Ni and NiO, however, with increasing O exposure the growth continues in an irregular way. We should mention that on the NiO crystallites atomic structures could so far not be resolved. More details of growth and analysis of NiO/Ni(100) are described in Ref. 15.

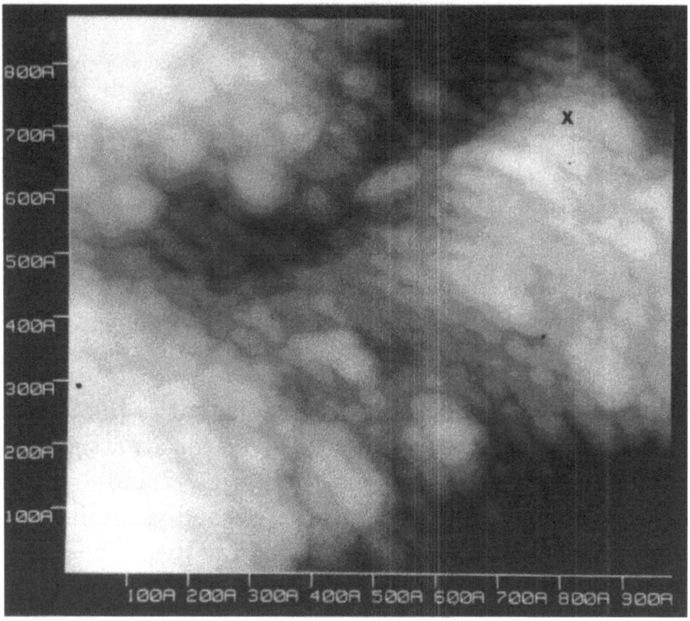

Fig. 9. CCT from NiO/Ni(100) obtained at $U = 0.5$ V and $I = 0.5$ nA. The range of gray tones corresponds to a corrugation of 60 Å. The individual NiO islands have a typical corrugation of 5...10 Å.

4. CONCLUSION

We have demonstrated the usefullness of STM for the analysis of a nonreactive (Ag/Ni(100)) and a reactive (O/Ni(100)) adsorbate/substrate system including the substrate itself. Topographical details of the substrate as terrace and step structures are found to play important roles for the behavior of the adsorbate. In the case of Ag condensation the steps act as preferential sites for the formation of two-dimensional Ag nuclei. Since the largest NiO clusters on a low-coverage O/Ni(100) surface are also observed at steps, we believe that the growth of "epitaxial" NiO also starts on these locations. The reason for the preferential occupation of step sites in case of Ag may simply be one of higher binding energy of Ag nuclei to the Ni substrate in comparison with a terrace site. For the NiO islands the differences of the stoichiometry between oxide and metal and the supply of Ni atoms from a step rather than from a terrace site may be the origin of preferential NiO growth at steps. A general result of our experiments was the complexity of structures found on the various surfaces. Although the corresponding LEED pattern always indicated a high degree of long-range order, the STM images indicated the presence of a more local order. The absence of a well-developed long-range order on the STM images may be characteristic for Ni(100), which is difficult to clean and contains residual contamination, which could influence the ordering of adsorbates. On the other hand, the possibility of measuring irregular and random structures is a main advantage of STM.

ACKNOWLEDGEMENT

This work has been supported by the Deutsche Forschungsgemeinschaft. We thank H. Wengelnik for preparation of the tunneling tips and D. Badt for technical help. We are indebted to G. Hackel for her photographic work.

REFERENCES

1 Neddermeyer H (1990) to be published in Crit Rev Solid State Mat Sci
2 Baro AM, Binnig G, Rohrer H, Gerber Ch, Stoll E, Baratoff A, Salvan F (1984) Phys Rev Lett 52:1304
3 Stroscio JA, Feenstra RM, Fein AP (1987) Phys Rev Lett 58:1668; Hamers, RJ, Avouris Ph, Boszo F (1987) Phys Rev Lett 59:2071
4 Wintterlin J, Wiechers J, Brune H, Gritsch T, Höfer H, Behm RJ (1989) Phys Rev Lett 62:59
5 Kittel C (1976) Introduction to Solid State Physics, John Wiley, New York/London/Sydney/Toronto
6 Shapiro AP, Wachs AL, Miller T, Chiang TC (1985) Solid State Commun 55:1101; Shapiro AP, Hsieh TC, Wachs AL, Miller T, Chiang TC (1988) Phys Rev B 38:7394
7 Daum W (1987) J Electr Microsc Rel Phen 44:271; Black JE, Mills DL, Daum W, Stuhlmann C, Ibach H (1989) Surface Sci 217:529
8 Ohtani H, Kao CT, van Hove MA, Somorjai GA (1986) Prog Surface Sci 23:155
9 Starke U, de Andres PL, Saldin DK, Heinz K, Pendry JB (1988) Phys Rev B 38:12227
10 Franchy R, Wuttig M, Ibach H (1989) Surface Sci 215:65
11 Bartelt NC, Roelofs LD, Einstein TL (1989) Surface Sci 221:L750
12 Saiki RS, Kaduwela AP, Osterwalder J, Fadley CS (1989) Phys Rev B 40:1586
13 Hono K, Iwata T, Pickering HW, Sakurai T (1989) Surface Sci 209:L109

14 Wang WD, Wu NJ, Thiel PA (1990) J Chem Phys 92:2025
15 Bäumer M, Cappus D, Kuhlenbeck H, Freund HJ, Wilhelmi G, Brodde A, Neddermeyer H (1990) to be published
16 Binnig G, Fuchs H, Stoll E (1986) Surface Sci 169:L295
17 Behm RJ (1990) private communication
18 Berghaus Th, Brodde A, Neddermeyer H, Tosch St (1987) Surface Sci 184:273
19 Wilson RJ, Chiang S (1987) Phys Rev Lett 59:2329
20 Tosch St, Neddermeyer, H (1988) J Microscopy 152:415
21 Witten TA, Sander LM (1983) Phys Rev B 27:5686

MOLECULAR IMAGING WITH THE SCANNING TUNNELING MICROSCOPE

Jürgen P. Rabe

Max–Planck–Institut für Polymerforschung
Postfach 3148, D–6500 Mainz, W–Germany

Abstract: Two classes of molecular materials can be directly investigated by scanning tunneling microscopy (STM), namely (1) organic conductors and (2) ultrathin organic layers on conducting substrates. Since STM itself does not require ultrahigh vacuum conditions, both the free surfaces as well as interfaces with fluid ambients are accessible. Unique is the ability of STM to investigate molecular defects on the atomic length scale. Moreover, it allows, for the first time, to directly observe molecular dynamics on the time scale of milliseconds and longer. After the description of some experimental aspects of fast STM imaging of molecular materials, a number of examples will be discussed in some detail. They include a radical cation salt, β–(BEDT–TTF$)_2I_3$, under ambient conditions and monolayers of an alkane and an alkyl–derivative (didodecylbenzene) adsorbed at the interface between an organic solution and the basal plane of graphite.

INTRODUCTION

The scanning tunneling microscope (STM) became famous due to its ability to image semiconductor surfaces under ultrahigh vacuum conditions at atomic resolution [1]. Later it was shown that the method itself does not require vacuum, but that atomically resolved images could also be obtained at various solid–fluid interfaces. This rendered STM particularly attractive for organic materials, which often are neither prepared nor employed under vacuum conditions. Reviews describing STM in general, including some applications to surface chemistry can be found elsewhere [2,3]. The basic requirements for a system to be investigated by STM are twofold: (1) sufficient electrical conductivity and (2) a limited mobility at its surface. Not required, on the other hand, is any particular degree of order. As far as molecular materials are concerned this means that two classes of systems lend themselves to STM studies: (1) Organic conductors and (2) monomolecular layers (thickness $d \lesssim 1$ nm) of organic insulators on conducting substrates. The fact that STM is non–destructive and yields truly local information on the sub–nanometer scale makes it a unique method to study molecular defects in organic crystals as well as structure and dynamics of only partially ordered phases.

METHOD AND INSTRUMENTATION

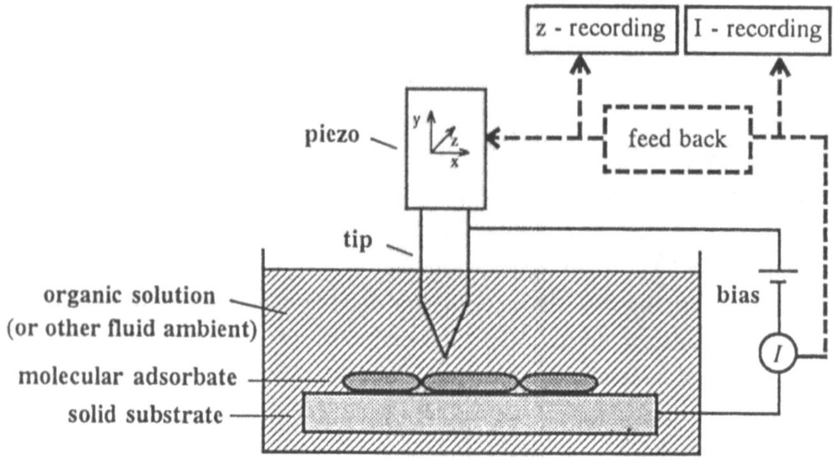

Fig. 1: In–situ scanning tunneling microscopy (STM) at the solid–fluid interface

The instrument employed here has been described before [4]. It was designed with an emphasis on (1) high scanning speeds (1 ms per line) and (2) low currents (down to 1 pA). The high speed has practical advantages (fast search as well as tolerance against thermal drift) and, moreover, it allows to follow some molecular surface dynamics. The low current is advantageous for non–destructive imaging of relatively poor conductors.

For the scanner a single piezoelectric tube (EBL Co., East Hartford, CT) with four segmented electrodes is used. It allows to scan the tip in all three directions in space across about 1.5 μm. Coarse approach of tip and sample is accomplished with a motor–driven differential micrometer. The STM head is placed on a stack of metal plates separated by viton spacers. It is mounted inside an evacuable bell jar which allows operation in controlled atmospheres. The bell jar together with an adsorption pump is placed on an optical table supported by an air spring vibration isolation system. All experiments have been performed at room temperature. The sensitivity of the current–voltage conversion is between 10^7 and 10^{10} V/A, allowing operation between 1 pA and 100 nA. All images presented here were recorded at about 1 nA with a tip bias around + 1 V. The STM electronics allows both recording of the tip height (constant current mode) and recording of the current (constant height mode). In the constant current mode scanning and data acquisition are accomplished with an IBM PC/AT. Faster and higher resolution images are obtained in the constant height mode. In this case the tip is scanned with up to 1 kHz in x–direction and 40 Hz in y–direction, implying a time resolution of 100 ms for an image of 100 lines, i.e. about real time to the eye. The tunneling current is converted on–line into a video signal by an image processor (Arlunya TF6540, Vermont, Australia) and continuously displayed, as well as stored on video tape. Sections of interest can be reviewed later and continuously averaged, or digitized for further image processing on a graphics workstation (Silicon Graphics). For the images presented here digital image processing was limited to the subtraction of tilt planes, the removal of some high frequency noise and to generate the colored images.

A crucial component in a STM is the tip. Its shape plays a decisive role for STM imaging of surfaces, which are rough on the nanometer scale. For atomically flat surfaces the general shape is less important, but still the contrast in an atomic resolution image is influenced by the exact atomic structure of the metallic cluster at the end of the tip. Unfortunately, this structure is difficult to characterize in–situ and, even more important, it is also not perfectly stable during scanning at room temperature. The work presented here has been carried out with Pt/Ir tips, which have been either electrochemically etched (ac etch in aqueous NaCN/KOH solution) or mechanically formed. Their quality was judged by the images they produced on a well known surface like the basal plane of highly oriented pyrolytic graphite. If resolution, contrast or stability were not as expected, a short bias pulse (e.g. 4 V, 100 ns) was applied to the tip during scanning. This altered usually the image contrast and is attributed to a change of the tip–structure. It can be associated with field evaporation from the tip and may also modify the sample at this position [5–7], which means that one has to move to another area of the sample. With this tip preparation, no significant differences were observed between tips which had initially been electrochemically or mechanically sharpened.

ORGANIC CONDUCTORS

A class of materials which has attracted considerable interest in recent years are organic conductors. In particular conjugated polymers have been investigated extensively, both in order to understand the electronic conduction mechanisms and also to possibly devise applications in electronics or optoelectronics. However, despite considerable efforts the understanding is still fairly limited. Part of the difficulty in obtaining reliable experimental data on the transport properties is the complicated morphology of the materials, which cause a poorly defined contact between polymer molecules or bundles with the external circuitry. In this respect the STM may open new ways since it offers the possibility to address and "interface" to selected areas on the sample. In order to demonstrate the capabilities of STM on organic conductors, a well defined crystalline material will be discussed first.

An organic conductor which can be readily imaged by STM under ambient conditions is the β–form of the iodine salt of bis(ethylenedithio)tetrathiafulvalene β–$(BEDT\text{–}TTF)_2 I_3$ [8]. The material is a radical cation salt, which is stable under ambient conditions. It is a normal conductor at room temperature and becomes superconducting below 1.6 K at ambient pressure. It crystallizes into a triclinic unit cell with well known parameters (a = 662 pm, b = 910 pm, c = 1529 pm, $\alpha = 94^0$, $\beta = 96^0$, $\gamma = 110^0$). In Fig. 2 STM images of the ab–plane of β–$(BEDT\text{–}TTF)_2 I_3$ are displayed. The low resolution image (Fig. 2a) shows flat terraces and steps which are about 1.5 nm high, corresponding to a molecular monolayer. It should be noted, however, that the stability of the image in air is not perfect, i.e. some material is removed during scanning. In fact, Fig. 2a shows a hole which had been formed during the consecutive recording of STM images. Clearly the damage becomes larger during scanning, indicating that it is not caused by pure chemical degradation. Similar observations had been made for another organic conductor, TTF–TCNQ [9]. A possible reason is field evaporation [5] of the molecules at the edges, which could be caused by an increased electric field at the edge or, alternatively, by some overshoot of the feed back, causing a reduced gap width for a short time, which in turn would also result in an increased electric field.

Nevertheless, high resolution images can still be continuously obtained even though occasionally a monolayer is scraped away, since on a crystalline material a fresh and identical surface is prepared during this procedure. Fig. 2b shows a high resolution image from the center of Fig. 2a, with the unit cell clearly visible. Also there is some more detailed structure which, however, is difficult to assign without any further consideration of the electronic structure of the material.

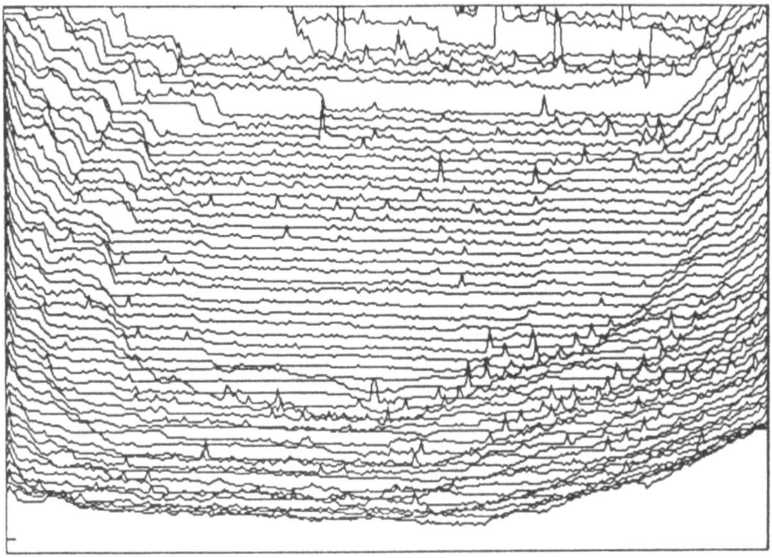

(a)

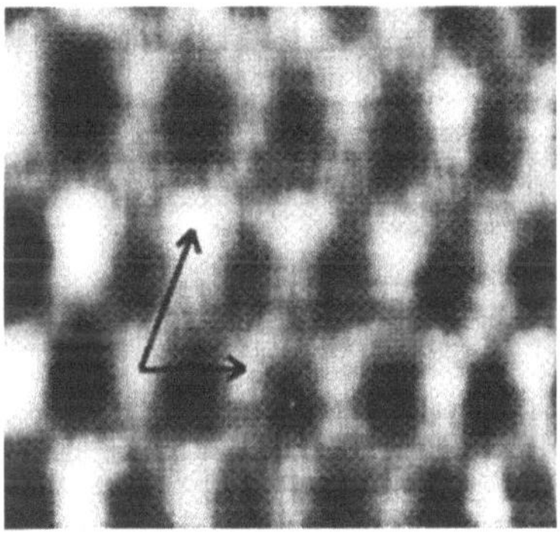

(b)

Fig. 2: STM images of the ab–plane of β–(BEDT–TTF)$_2$ I$_3$ [8].
(a) Quasi 3–dimensional display of a 'Constant Current' image. The image size is 100 nm x 25 nm and the mark at the lower left corner corresponds to 1 nm in height.
(b) Top view 'Constant Height' image with the brightness proportional to the current. The image size is 3.5 nm x 3.2 nm.

Iodine doped polyacetylene has also been imaged by STM [8]. In this case it was necessary to first compress the sample into a more compact film than initially prepared, in order to obtain stable images. After that the images exhibit regions with parallel stripes, 5 nm to 10 nm in width, which may be attributable to bundles of polymers, indicating that the STM provides a means to address single polyacetylene fibrils with a metallic tip. In other words, STM may allow to interface an individual fibril of a conducting polymer to the external electronic circuitry.

MOLECULAR MONOLAYERS

STM imaging of organic molecules requires their immobilization on the surface of a conducting substrate. Molecular conformation as well as structure and electronic properties of a molecular monolayer, however, will strongly depend on the surface properties of the underlying substrate. This includes its chemical composition at the surface as well as its surface morphology on a molecular length scale. Unambiguous interpretation of STM images of a molecular layer requires, therefore, to well control and also characterize the substrate surface, preferentially also by STM.

Substrates

For high resolution STM imaging of an organic adsorbate the substrate surface should be as flat and regular as possible on a molecular scale. Under ultrahigh vacuum conditions single crystals of doped silicon or reactive metals are obvious choices. However, while a reactive surface may help to immobilize the adsorbate, it should be kept in mind that a strong interaction between substrate and adsorbate also implies that STM yields the interfacial structure rather than the image of the native molecule. Examples for this type of "chemisorbed" systems are ammonia on Si (111) [10] or benzene and carbonmonoxide coadsorbed on Rh(111) [11].

Less reactive substrates have advantages if (1) one wishes to retain more of the structure and properties of the unperturbed substrates or molecules or (2) one wants to study interfaces outside ultrahigh vacuum, where materials which readily form well insulating surface oxides, like silicon or aluminum, are obviously not suitable as substrates. More promising candidates in this case are the noble metals.

Gold and silver films of about 100 nm thickness thermally evaporated onto glass slides are commonly used as substrates for Langmuir–Blodgett (L–B) films or self–assembled monolayers. They appear flat in a Scanning Electron Microscope (SEM) or interference contrast microscope. However, STM reveals a grainy structure with grain sizes on the order of 25 nm [7,12]. The rms roughness of such gold films amounts to about 3.5 nm (or 30 nm peak–to–valley) over a typical scan range of 0.25 μm x 0.25 μm. This substrate roughness is fairly large if compared to molecular dimensions. Silver films are even a little rougher (5.5 nm rms and 35 nm peak–to–valley). Somewhat smoother but still rough on a nanometer scale are thin evaporation films of gold/palladium, carbon or platinum/carbon films on glass. However, also for these films the roughness appears to be too large for reliable molecular imaging of adsorbed monolayers with the STM.

Gold and silver films can also be evaporated onto mica. Films evaporated at a substrate temperature of 50^{0}C exhibit very similar grain diameters but flatter grains if compared to the films on glass: 1.2 nm and 1.9 nm rms (7.5 nm and 9 nm peak–to–valley) roughness for gold and silver, respectively. At higher temperatures (about 400^{0}C for gold and 275^{0}C for silver) the films grow epitaxially, giving rise to LEED patterns which correspond to a (111) surface. However, the STM images clearly show that atomically flat terraces of up to 100 nm in diameter coexist with ridges and holes, 10 nm or more in depth [7,12].

In summary, while gold and silver films evaporated at room temperature are too rough to allow unambiguous high resolution STM work on molecular adsorbates, films evaporated at elevated temperatures onto mica exhibit at least 100 nm terraces, which are sufficiently flat.

A complication which has not been addressed so far results from the fact that the surface of a metal or semiconductor will in general reconstruct under the influence of an adsorbate. Only substrates whose surface structure is extremely stable can be expected to remain unperturbed.

An ideal substrate, in many respects, is highly oriented pyrolytic graphite (HOPG). Its surface is atomically flat and perfect over distances on the order of a micrometer and, moreover, it is mechanically very stable and chemically inert, which means in particular that it does not readily form any insulating surface layer. The fact that it can be imaged at atomic resolution in a variety of fluid ambients indicates that the mobility of the ambient adsorbates is so high that they appear on the images as noise only. On the other hand, this raises the question how to immobilize an adsorbate on an inert substrate.

In a first attempt to image an organic monolayer by STM a fatty acid salt has been transferred to HOPG by the L–B technique [13]. While STM images consistent with an ordered monolayer could be obtained, they were limited to very small parts of the sample. A difficulty was the poor transfer of fatty acid L–B films on HOPG. Subsequent characterization of the transferred films by vibrational spectroscopies showed that the structure of the transferred monolayer is significantly different from the film structure on the water surface or other common L–B substrates like glass, silicon or aluminum [4]. In particular, grazing–incidence reflection infrared spectroscopy showed that the alkyl chains are oriented parallel to the substrate rather than perpendicular. This indicates that (1) alkyl chains tend to lie flat on HOPG and (2) it was not possible to force them by the L–B method into a stable different structure. Attempts to prepare inherently more stable L–B monolayers from stiff polymers are promising, but the results obtained so far indicate again substantial inhomogeneities on the molecular scale [4,6].

Another class of organic systems which were studied by STM were liquid crystals on HOPG [14,15]. STM images indicate, however, that the first monolayer at the interface with HOPG exhibits a crystalline structure with the molecules oriented parallel to the graphite plane. Similarly do long chain alkanes adsorb as highly ordered monolayers of flat lying lamellae, as evidenced by both adsorption isotherms and STM [16–18].

It should be noted that the alkyl derivatives described above were always oriented parallel to HOPG, regardless whether they were applied as L–B film, liquid crystal or adsorbed from solution, indicating that this conformation is thermodynamically stable.

Alkanes on HOPG

Organic monolayers of long chain alkanes on HOPG have been formed by direct adsorption from an organic solution. Although this adsorption is known for some time, detailed structural information on such layers in–situ is scarce if compared, for example, to monolayers preformed at the water–air interface and transferred subsequently to a solid support by the L–B technique. Part of the reason is the lack of experimental methods which allow to determine the molecular structure at the solid–fluid interface.

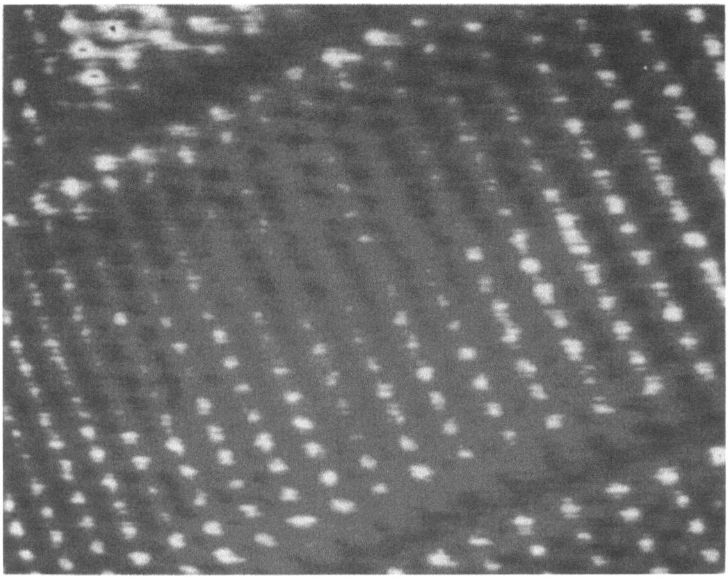

Fig. 3: STM image of a dotriacontane ($C_{32}H_{66}$) monolayer on HOPG.

Fig. 3 exhibits a STM image of a monolayer of the long chain alkane dotriacontane ($C_{32}H_{66}$) on the basal plane of graphite. The monolayer has been adsorbed from a solution in phenyloctane (concentration: 2 mg/ml) and was imaged in–situ. The image shows a lamella of extended flat lying molecules. Individual chains are resolved with a separation between them of 0.43 nm and the repeat unit along the chain of 0.25 nm, corresponding to two methylenes. The molecules are oriented perpendicular to a lamella boundary and from lamella to lamella a displacement by half a molecular spacing is observed. Similar images have been obtained for other long–chain alkanes.

In order to address the question whether any molecular dynamics can be observed on the time scale accessible here, we have continuously recorded STM images of the center part of a dotriacontane lamella. While an image like the one given in Fig. 3 may be stable on a time scale of many seconds or minutes, it happens that defects involving individual molecules occur, which then may diffuse through the lamellae or heal out again [20]. Fig. 4 shows a snap shot of a video recording, displaying the occurrence of such a defect.

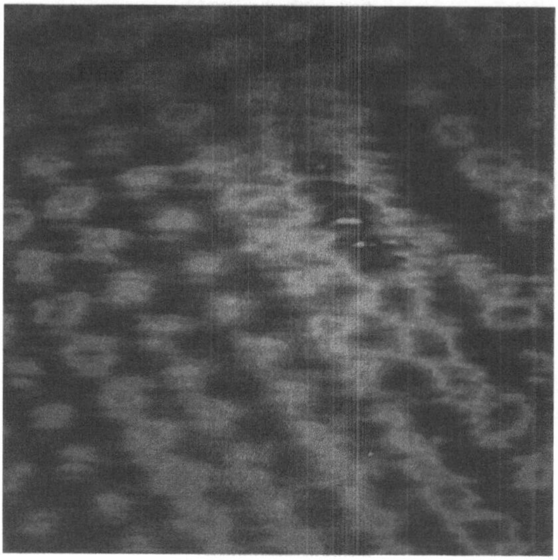

Fig. 4: Snap shot of STM video–recording, showing the center of a lamella of the alkane dotriacontane adsorbed at the interface between an organic solution and HOPG. The lower left part exhibits parallel lying molecules. On the upper right a defect is visible. It is stable for only about a second. Image size: 3.0 nm x 3.0 nm.

Didodecylbenzene on HOPG

Besides long chain alkanes also a number of alkyl derivatives form lamellar monolayers, which similarly can be imaged in–situ by STM. One example is didodecylbenzene ($H_{25}C_{12}(C_6H_4)C_{12}H_{25}$ or DDB for short) adsorbed to HOPG from concentrated but not saturated solutions (concentration: 10 mg/ml) in phenyloctane, as already described elsewhere [19,20]. Fig. 5 shows STM images exhibiting two domains which are separated by a molecularly sharp boundary. Within the domains lamellae with equal width are observed. The contrast at the position of the benzenes is dominating, but the alkyl sidechains can also be resolved. At the domain boundary the lamellae are displaced and their orientation is rotated within the plane by 60°. This type of domain boundary is typical for the monolayer. It is so abundant that hardly any lamella is longer than about 30 molecules. Fig. 6 shows the molecular model explaining the observed images: The

(a)

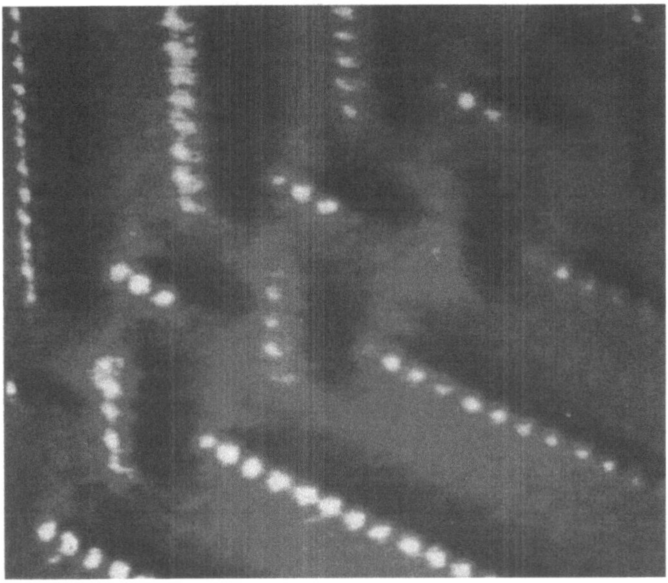

(b)

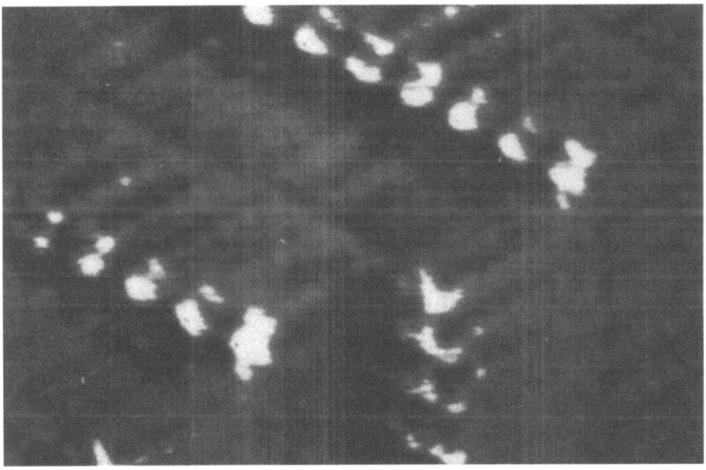

Fig. 5: STM images of a monolayer of didodecylbenzene adsorbed at the interface between an organic solution and the basal plane of graphite [20]. The images show a 2–dimensional polycrystal with two kinds of lamellar domains separated by a characteristic domain boundary. (a) The strongest contrast results from the phenyl stacks. (b) Under appropriate tip conditions the sidechains become also visible. Evidently they are tilted relatively to the lamella boundary and parallel to each other in adjacent domains. Image size: (a) 13.3 nm x 11.2 nm; (b) 7.2 nm x 4.7 nm.

DDB molecules are extended, exhibiting their lowest energy conformation, and lie flat on the HOPG. They pack in lamellae with the long axis of the alkyl side chains forming an angle of 60° with the lamella boundary. The domain boundary is a consequence of the symmetry of the molecule. Due to the displacement between the two alkyl sidechains at the benzene position, a rotation along the long axis of an alkyl chain leads to a symmetrically inequivalent orientation of the molecule relative to the substrate. Two adjacent domains consist of lamellae with the molecules in identical conformations but in the two different orientations. The sharp domain boundary between them is a screw axis. Fig. 6 shows that with this particular packing no major holes occur at the domain boundary. Therefore, the energy associated with this defect is small and it becomes understandable why these domain boundaries are so abundant.

Fig. 6: Molecular packing model for DDB monolayers on HOPG. Shown are three domains. The two different kinds of domains as well as the packing at the domain boundary are a direct consequence of the molecular symmetry. The center molecule is displayed without van der Waals radii, in order to display the molecular skeleton more clearly on the substrate.

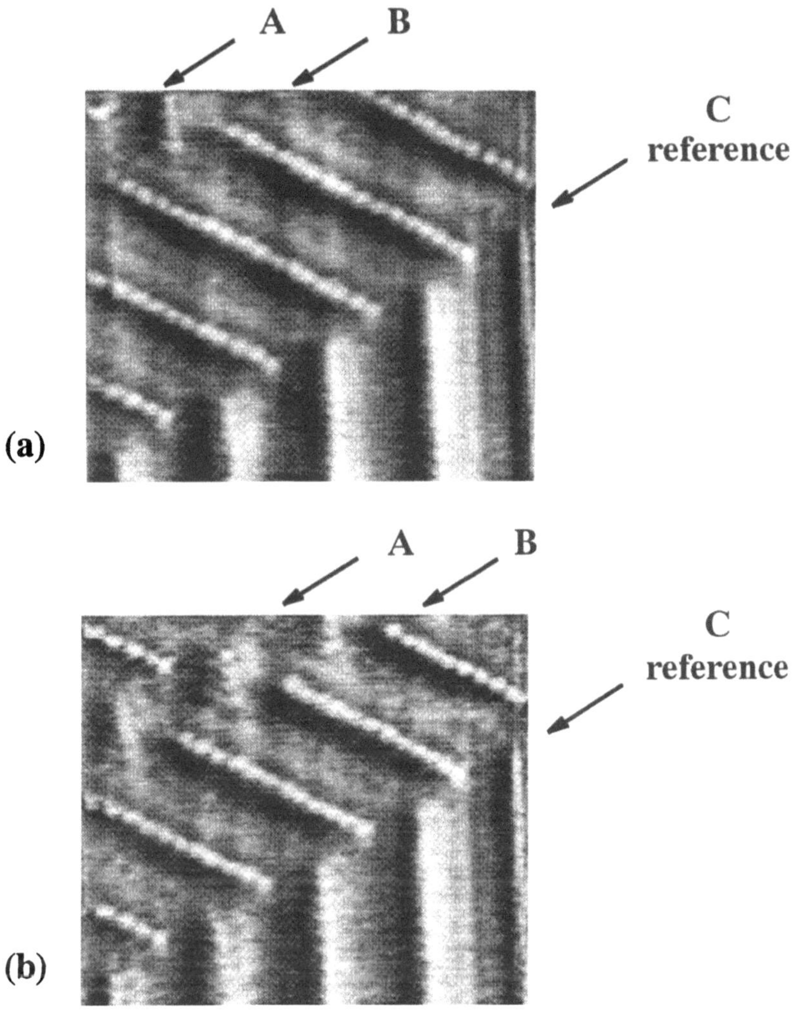

Fig. 7: Motion of domain boundaries in didodecylbenzene (DDB) on HOPG, illustrated by snap shots (a) and (b) of a video tape recording of STM images in "real time" to the eye [20]. While the lower domain boundary (C) remains fixed in time, the upper ones (A and B) move inward. The time between the frames is on the order of 10 sec. Image size: 14.5 nm x 12.0 nm.

The motion of such a domain boundary should to a first approximation cost no net energy. It only requires sufficient thermal energy to allow the cooperative motion of the molecules involved. The net molecular motion involves the rotation around the long molecular axis and the lateral displacement of the molecule by half the molecular length. The images Figs. 7a & b document the direct observation of such a process by STM. While the lower domain boundary, C, in Fig. 7a remains fixed in space for minutes, the upper domain boundaries, A and B, move inward (b) on the time scale of seconds. The lamellae between B and C shrink first by four molecules and expand then again. Since at least three lamellae were involved this means that at least 12 molecules had to move cooperatively in this domain.

Fig. 8 displays a series of snap shots for a case where a "free volume" diffuses through the crystalline monolayer. Initially a densely packed monolayer with two domain boundaries of the type given in Fig. 6 is observed (Fig. 8a). Then one lamella in the middle domain is extended by one molecule, for example by the shift–rotation of one molecule in the adjacent domain. As a consequence there is not enough space now for the lamellae in the adjacent domain to pack properly (Fig. 8b). The result is a "free volume" in the monolayer, probably filled by more mobile solvent molecules. Evidently, this situation is not very stable, and the defect diffuses around on the time scale of seconds. Given in Fig. 9 is the model for the free volume in Fig. 8b.

The stable images discussed so far have been obtained on highly ordered "nanocrystalline" domains. An interesting question is whether more disordered "glassy" phases can also be imaged. That this is indeed the case is demonstrated in Fig. 10, displaying a defect within a lamella of DDB on HOPG. Interestingly, the stack of phenyls appears unperturbed, while the alkyl sidechains around the defect are disordered. Contrary to the defect shown in Fig. 4 for the alkane dotriacontane the dynamics here was slow, i.e. the image was stable on the time scale of many seconds.

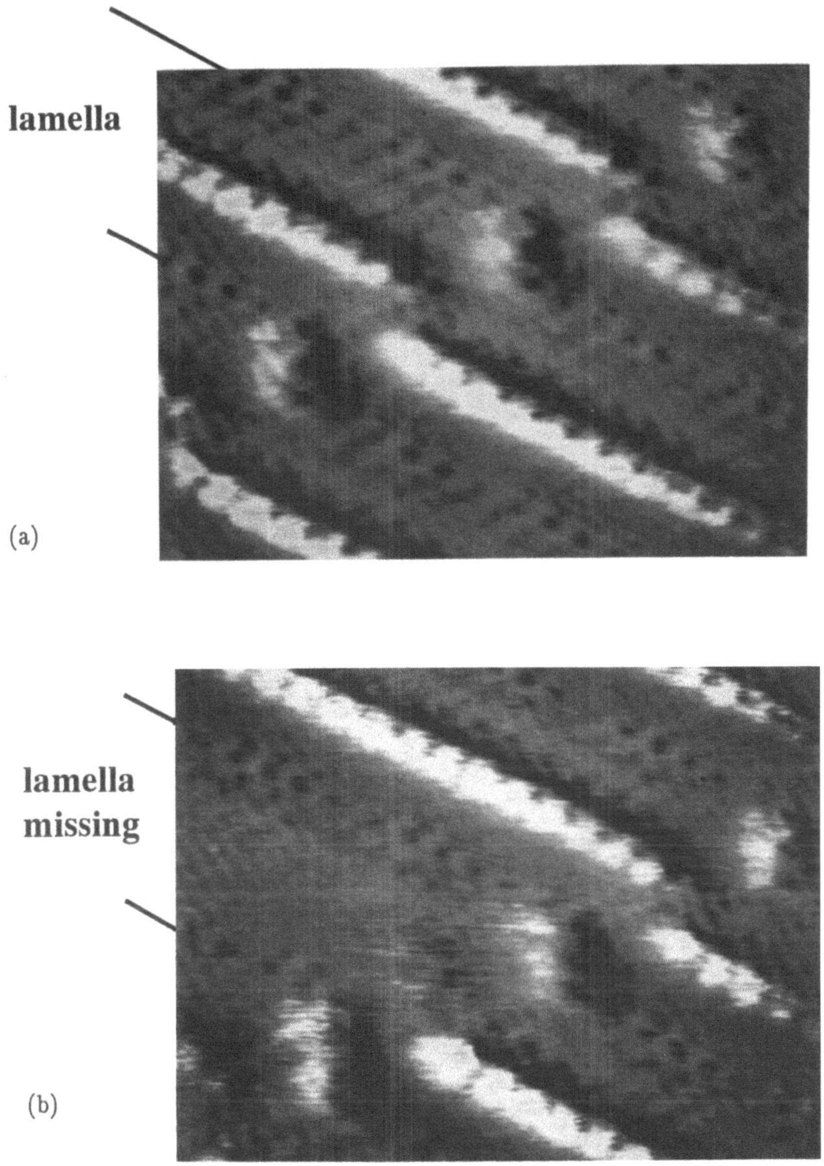

Fig. 8: Cooperative molecular motion around packing defect, illustrated by snap shots of a video recording of STM images [20]. (a) shows a densely packed monolayer with two domain boundaries of the type described in Figs. 5 & 6. In (b) one of the three lamellae in the middle domain has grown by one molecule, which does not allow the adjacent lamella in the upper domain to pack properly. The molecular model for this structure is given in Fig. 9. The "free volume" in the two–dimensional layer can easily diffuse on the time scale of seconds. Image size: 10.0 nm x 7.5 nm.

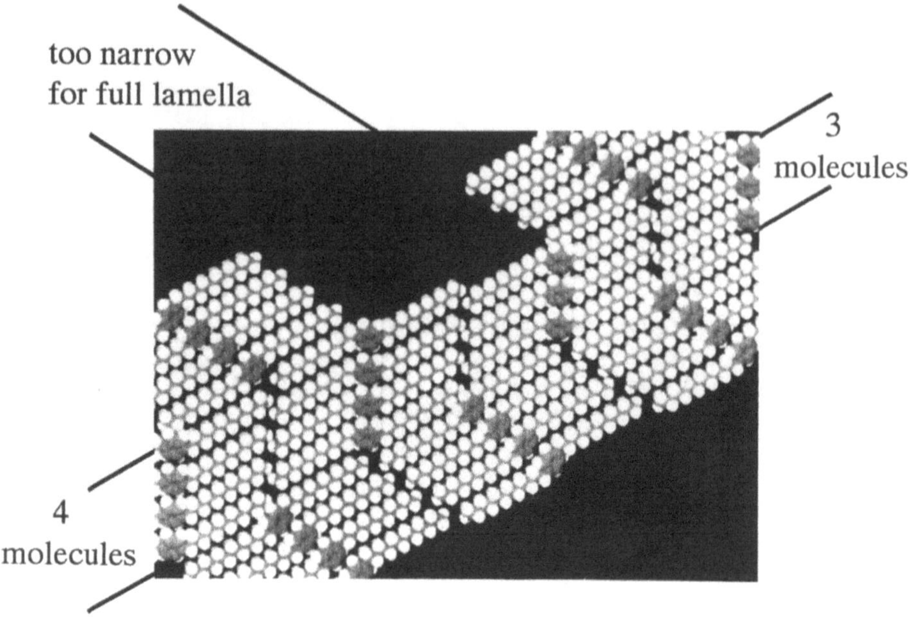

Fig. 9: Molecular model of the packing defect in Fig. 8b [20].

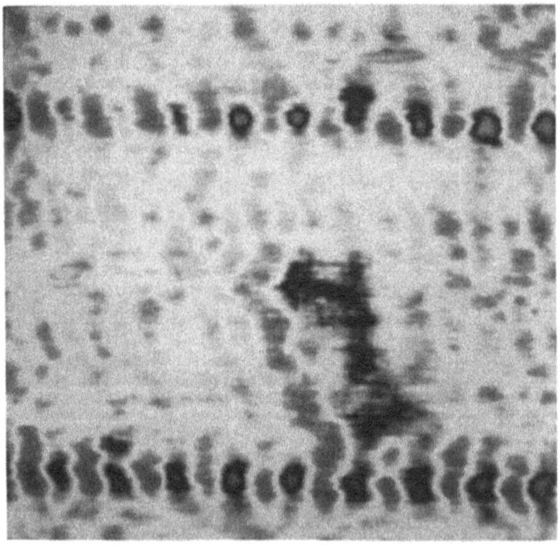

Fig. 10: Metastable defect and disordered sidechains in a lamellar phase of DDB on HOPG [19]. Image size: 5.0 nm x 4.7 nm.

SUMMARY

It has been demonstrated that scanning tunneling microscopy can be employed to investigate interfaces with molecular materials on the length scale of 0.1 nm and the time scale of 100 ms. This opens the possibility to address organic molecules with the STM tip, a prerequisite for any sort of spectroscopy or modification with the STM. The images obtained on organic conductors demonstrate a resolution better than the unit cell. The work on a conducting polymer with its more complicated morphology indicates that the STM may offer a way to form a well defined "interface" between the organic conductor and the external circuitry.

In–situ STM at the solid–fluid interface has been employed to investigate structure and dynamics of a long chain alkane, as well as an alkyl–derivative (didodecylbenzene), adsorbed as a monolayer from organic solution to the surface of highly oriented pyrolytic graphite. It was shown that in both cases the monolayers exhibit lamellar phases in which the molecules are extended and oriented parallel to the basal plane of graphite. While in the alkane the molecules are oriented perpendicular to the lamella boundary, in DDB they are tilted. Due to the symmetry of the DDB molecules two kinds of domains exist, which form molecularly well defined domain boundaries. A strength of STM is the ability to obtain truly local information. Defects involving individual molecules, as well as disordered glassy phases have been imaged. Finally, fast image recording allowed, for the first time, to directly observe molecular dynamics in an organic monolayer. Examples are the appearance, diffusion and annealing of defects involving individual molecules as well as the movement of domain boundaries on the time scale of 100 ms and longer. Clearly, STM is unique in determining both structure and dynamics in nanocrystalline or partly disordered monolayers on the sub–molecular scale.

ACKNOWLEDGEMENT

It is pleasure to acknowledge the contributions of S. Buchholz to the material covered in this review. The project has been supported by the Bundesministerium für Forschung und Technologie under the title "Ultrathin Polymer Layers" 03M4008E9 and by the European Science Foundation (Additional Activity: Chemistry and Physics of Polymer Surfaces and Interfaces).

1 Binnig G, Rohrer H, Gerber C, Weibel E (1983) Phys Rev Lett 50:120

2 Kuk Y, Silverman PJ (1989) Rev Sci Instrum 60:165

3 Rabe JP (1989) Adv Mater 1:13; Angew Chem Int Ed Engl Adv Mater 28:117; Angew Chem Adv Mater 101:117

4 Rabe JP, Sano M, Batchelder D, Kalatchev AA (1988) J Microscopy (Oxford) 152:573

5 Rabe JP, Buchholz S (1991) Appl Phys Lett 58:702

6 Rabe JP, Buchholz S, Ritcey AM (1990) J Vac Sci Technol A 8:679

7 Buchholz S, Fuchs H, Rabe JP (1991) Adv Mater 3:51

8 Rabe JP, Buchholz S (1990) In: Brédas JL, Chance RR (eds) Conjugated Polymeric Materials: Opportunities in Electronics, Optoelectronics, and Molecular Electronics. NATO–ARW Series E, Kluwer, Dordrecht, p. 483

9 Sleator T, Tycko R (1988) Phys Rev Lett 60:1418

10 Wolkow R, Avouris PH (1988) Phys Rev Lett 60:1049

11 Ohtani H, Wilson RJ, Chiang S, Mate CM (1988) Phys Rev Lett 60:2398

12 Buchholz S, Fuchs H, Rabe JP (1991) J Vac Sci Technol B 9:857

13 Smith DPE, Bryant A, Quate CF, Rabe JP, Gerber C, Swalen JP (1987) Proc Natl Acad Sci USA 84:969

14 Foster JS, Frommer JE (1988) Nature 333:542

15 Smith DPE, Hörber JKH, Binnig G, Nejoh H (1990) Nature 344:641

16 Groszek AJ (1962) Nature 196:531; (1964) Nature 204:680; (1970) Proc R Soc (London) a 314:473

17 Findenegg GH, Lippard M (1987) Carbon 25:119

18 McGonigal GC, Bernhardt RH, Thomson DJ (1990) Appl Phys Lett 57:1

19 Buchholz S, Rabe JP (1991) J Vac Sci Technol B 9, in press

20 Rabe JP, Buchholz S (1991) Phys Rev Lett 66:2096

IMAGING OF MAGNETIC DOMAINS IN FERROMAGNETS AND SUPERCONDUCTORS BY FORCE AND TUNNELING MICROSCOPY

U. Hartmann, R. Berthe, T. Göddenhenrich, H. Lemke, and C. Heiden

Institute of Thin Film and Ion Technology, KFA-Jülich,
P.O. Box 1913, D-5170 Jülich, Fed. Rep. Germany

Abstract: Near-surface variations of magnetic microfields have been observed by room-temperature magnetic force microscopy (MFM) on ferromagnetic media. Apart from presenting data of current importance to basic micromagnetic research, fundamentals of contrast formation in MFM are discussed in detail. Using low-temperature scanning tunneling spectroscopy (STS) the spatial arrangement of magnetic domains in a type-II superconducting medium has been directly imaged. First promising STS results, obtained by employing ferromagnetic tips for tunneling, provide an important basis for future applications of low-temperature MFM to the analysis of superconducting media.

INTRODUCTION

Ferromagnets and superconductors are of outstanding importance to numerous present-day technical applications. However, in future new philosophies for microelectronic devices, predominantly in the fields of improved digital data storage and processing concepts and microsensor development, will further strongly increase the need for optimized ferromagnetic and superconducting materials. A common key feature of these two types of ordered media is their detailed response to external magnetic fields.

For ferromagnets the macroscopic magnetic properties are controlled by the collective behavior of an ordered network of microscopic magnetic domains. Magnetization reversal proceeds by motion of interdomain boundaries and by nucleation and annihilation of domains. The detailed characterization of the most basic phenomena in ferromagnetism, fundamental to the development of elaborate magnetic microdevices, strongly relies on improved techniques for imaging magnetic domain structures at high spatial resolution.

Magnetic force microscopy (MFM) [1,2] utilizes a fine ferromagnetic probe to sense near-surface microfields emenating from a ferromagnetic sample. As an offspring of the atomic force microscope (AFM) [3], the instrument permits the detection of minute probe-sample forces at a sensitivity somewhere in the pN range.

MFM has been used previously to image a variety of ferromagnetic media. Natural domain configurations in NdFeB [4-6], CoCr [6,7], TbFeCo [8], Permalloy [9], and Fe single crystals [10-12] have been identified with sub-100nm resolution. Intentionally written magnetization patterns have been studied on a number of recording materials, including Co-alloy thin films [13-15], γ-Fe_2O_3 particulate media [15], and TbFe [14,16], a material for magneto-optical recording. The high-resolution analysis of thin film magnetic recording heads [1,17] can also be considered as an MFM application of considerable technical relevance.

With respect to present-day applications of superconductivity, type-II superconducting media are of major technical importance. If these media are subject to a magnetic field, the flux penetrates the medium in terms of an ordered arrangement of cylindrical magnetic domains. These vortices exhibit a normal conducting core region. The macroscopic electronic properties of type-II superconductors, fundamental to the detailed behavior of superconducting devices, are largely controlled by the stability and dynamics of the field-induced vortex lattice. Imaging of the latter provides a fundamental basis for superconducting material optimization and device development.

Scanning tunneling spectroscopy (STS) probes the local density of electronic surface states utilizing the high spatial resolution of scanning tunneling microscopy (STM) [18]. First applications of low-temperature STS to superconductors clearly confirm the capability of imaging local variations in superconductivity at a nm scale [19-21].

As stated above, the present analysis of magnetic domains in ferromagnets and superconductors involves force microscopy for the former type of material and methods of tunneling microscopy for the latter. However, the exclusive applicability of either of these scanned probe techniques to magnetic imaging is likely to be removed in future. On the one hand, first direct STM observations of magnetic phenomena via spin-polarized vacuum tunneling have been reported recently for the system CrO_2/Cr(001) [22]. On the other hand, recent experiments in our laboratory strongly support the possibility of detecting vortices in type-II superconductors by low-temperature MFM.

MAGNETIC FORCE MICROSCOPY ON FERROMAGNETS

Principles of Contrast Formation

MFM relies on the spatially resolved mapping of minute magnetostatic forces (or force gradients) exerted on a fine ferromagnetic probe, subject to the near-surface microfield configuration of the magnetic sample . The probe-sample spacing is typically between 10nm and more than 1μm. Since magnetic forces can either be attractive, repulsive, or absent, an overall attractive servo force is generally superimposed by applying a voltage (typically 1-10V) between probe and sample to ensure proper operation of the feedback system. The Coulomb force is given by

$$F_c = (V^2/2)\ \partial C/\partial d \quad , \tag{1}$$

where V is the applied voltage and C the probe-sample capacitance which depends on the probe-sample spacing d. Adjustment of the mean working distance is straightforward by choosing an appropriate servo voltage. It has recently been demonstrated that the application of an ac-modulated servo voltage additionally allows a complete separation of topographic and magnetic features of the sample which generally both contribute to the MFM image [17].

As shown in Fig.1, the stray field H(r) of a ferromagnetic sample originates from divergences of the near-surface magnetization vector field M(r′) and from surface charging due to **M** components locally oriented along the axis of the surface normal vector s:

$$H(r) = \frac{1}{4\pi}\left[-\int \nabla\cdot M(r')\ \frac{r-r'}{|r-r'|^3}\, d^3r' + \oint \frac{r-r'}{|r-r'|^3}\, M(r')\cdot ds \right] \quad . \tag{2}$$

The internal magnetic structure of the sharp microprobe is determined by the detailed balance of exchange, crystalline anisotropy, and magnetostatic energy contributions. This balance generally results in a complex multidomain configuration as shown in Fig.1. Since magnetic forces are long range in nature the overall tip magnetic configuration is *a priori* relevant to MFM (unlike the case for STM and repulsive-mode AFM). However, sufficiently far away from the tip apex the domains arrange in such a way that a largely stray-field-free magnetic flux closure is obtained for reasons of energy minimization. On the other hand, in vicinity to the apex, total flux conservation within the tip is no longer energetically favourable. Due to the enormous shape anisotropy the magnetization is predominantly oriented parallel to the probe axis. A complex arrangement of magnetic surface and bulk charges is created which forms the "active detector volume" of the probe.

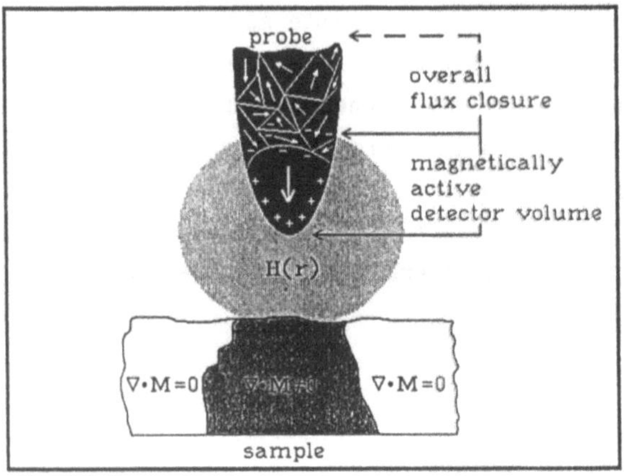

Fig.1.
Magnetostatic fundamentals of contrast formation in MFM. Bulk and surface divergences of the sample magnetization vector field **M** produce a highly inhomogeneous near-surface microfield H(r) which interacts with the complex internal domain configuration of the probe. The active detector volume of the latter consists in a magnetically charged region in close vicinity to the geometrical apex.

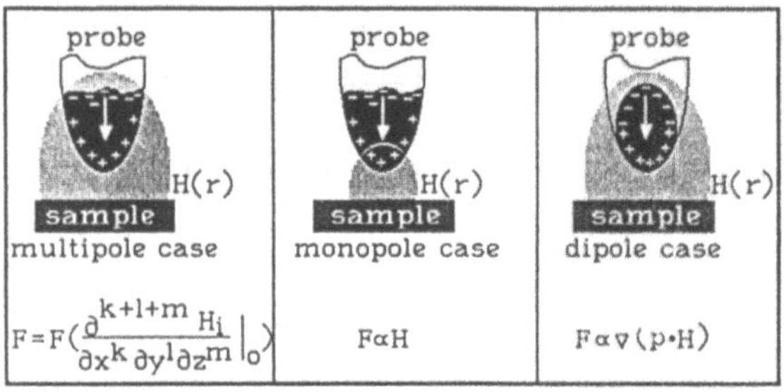

Fig.2.
Dependence of the magnetostatic force F(r), as detected by MFM, on the sample microfield H(r). For the multipole case F is determined by all derivatives of H with respect to the apex coordinates. For the monopole case F is simply proportional to H and for the dipole case (moment p) F is determined by the first-order derivatives of H.

Upon immersion into the sample stray field, the magnetostatic force acting on the probe is generally determined by the multipole interaction shown in the left column of Fig.2. In this complex case, image interpretation is difficult since the obtained contrast is determined by several multipole components of the probe and higher-order derivatives of the sample microfield H(r). The situation considerably simplifies if we have an elongated homogeneously magnetized apex domain within the tip and a comparatively short-range sample stray field (middle column of Fig.2). In this case the probe behaves like a magnetic monopole and the MFM directly detects H(r). The monopole case is realized, e.g., for the detection of narrow domain boundaries. On the other hand, if the active detector volume of the probe consists in a small spherical (or slightly elongated) single domain particle, and if the sample microfield is long range compared to the dimensions of this particle (right column of Fig.2), the probe only exhibits a dipole component and the MFM detects the first-order derivatives of **H**(**r**). The latter case is frequently realized for the detection of an extended magnetic structure, intentionally written into a recording medium [24-26]. The art of preparing suitable MFM probes is thus to create monopole or dipole probes rather than to fabricate extremely sharp tips.

With respect to sharp tips there is another point which deals with the question of destructiveness of MFM. Due to magnetic flux escape, the probe itself creates a stray field of considerable magnitude. For extremly soft magnetic media the sample magnetization may thus be deformed [9,10,27]. The destructive self-magnetic fields have their maxima in close vicinity to sharp corners. To demonstrate this phenomenon, we intentionally prepared Ni tips exhibiting a large number of sharp structures. Figure 3 shows the typical stray field configuration resulting from such a serrated model tip.

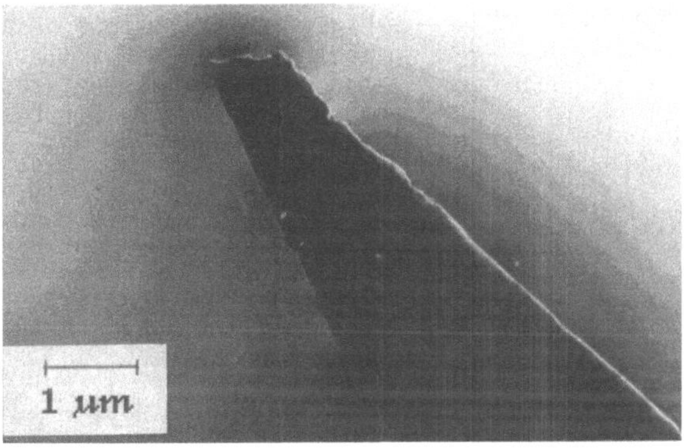

Fig.3. Foucault mode Lorentz micrograph of a serrated Ni tip. Dark and light patterns outside of the tip are due to the stray magnetic field component parallel to the tip axis. (Courtesy of J. Chapman and S. McVitie, University of Glasgow)

Figure 4 presents theoretical data obtained according to Eq.(2) for a probe which would, at first sight, be considered as well suited for high-resolution MFM: a sharp homogeneously magnetized cone. However, the calculations confirm the strong field-focussing effect resulting from the conical shape. The stray fields clearly exceed the saturation magnetization of the tip material. For realistic cone angles, the fringe field reaches its maximum right at a distance from the apex which corresponds to the length of the apex domain as shown in Fig.4(a). Upon decreasing the distance, the field first decreases rapidly and then again increases reaching a logarithmic singularity at the tip apex. For distances from the apex which exceed the tip domain length, the field then exhibits a monotonic decrease. For half-cone angles larger than 40^{0} the stray field of the tip exhibits a solely monotonic decrease with increasing distance from the apex. The stray field as a function of the cone angle is shown in Fig.4(b) for some characteristic distances from the apex.

For distances smaller than the tip domain length, the field maximum is obtained at a half-cone angle of about 50^0. At a distance from the apex which corresponds exactly to the domain length, the stray field of the cone strongly increases with decreasing cone angle. Finally, at distances well above the tip domain length of the cone, the stray field is practically independent of the cone angle.

The above results show that the response function in MFM, which determines such basic fundamentals as force sensitivity, spatial resolution, and sensed field component(s) or derivative(s), does not simply depend on the overall probe geometry, but is determined by the magnetic configuration in vicinity to the probe apex. MFM probes should have smooth surfaces to avoid destructive stray fields which can easily exceed the probe saturation magnetization.

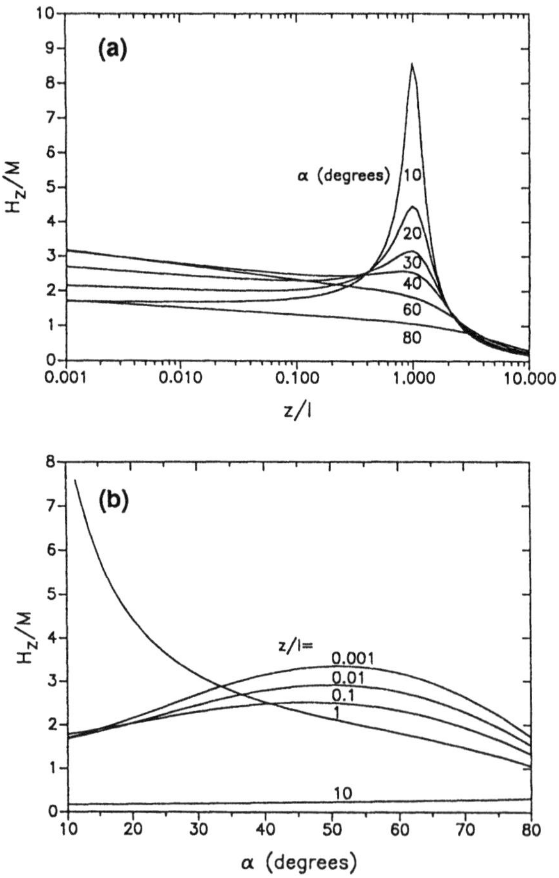

Fig.4.
Magnetic stray field of a conical MFM tip with a homogeneously magnetized apex domain. M denotes the tip magnetization, α the half-cone angle, and l the domain length. (a) Dependence of the field on the axial distance to the tip apex. (b) Dependence of the field on the half-cone angle.

Van der Waals Forces

Apart from electrostatic and magnetostatic forces the long-range van der Waals (vdW) forces provide an ever-present contribution to probe-sample interaction in MFM. We recently presented a general theory on vdW interactions which is rigorously based on principles of macroscopic quantum field theory [28,29]. As a major result, the interaction comes out to be completely nonretarded at small probe-sample distances (in the nm range), while retardation effects of the radiation field between probe and sample (Casimir effect) become noticeable for sufficiently large probe-sample spacings (in the 10nm range). For the limiting regimes we found simple inverse power laws of type

$$F(d) = \begin{cases} g_n\,(H_e+H_n)\,d^{-n} & \text{at small distances} \\ (g_nH_e+g_rH_rd^{-1})\,d^{-n} & \text{at large distances} \end{cases} \qquad (3)$$

where the geometrical factors g_n and g_r as well as the index n depend on the probe geometry (Fig.5). The dielectric properties involved are incorporated into the entropic, nonretarded, and retarded Hamaker constants H_e, H_n, and H_r.

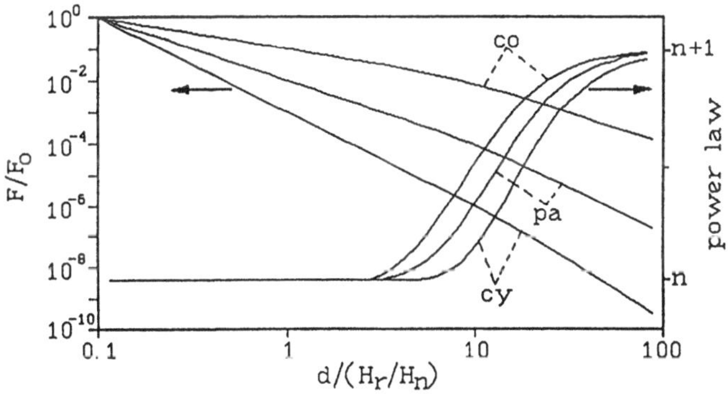

Fig.5.
vdW forces as a function of probe-sample spacing in vacuum for typical metals. The probe is approximated by a cylinder (cy), a paraboloid (pa), and a cone (co), respectively. Upon onset of retardation the power law index increases from n to n+1, with n=3 (cy), n=2 (pa), and n=1 (co).

Figure 6 shows the effect of some highly polar H-bonding immersion media which are known to lower vdW forces with respect to vacuum. The most striking feature of these results is the postulated crossover from attractive to repulsive forces for H_2O, H_2O_2, and glycol at large distances (inaccessible to force microscopy), and the repulsive interaction for glycerol and formamide at any distance. The parasitic vdW background in MFM can thus be handled to a certain degree by choosing an appropriate liquid environment.

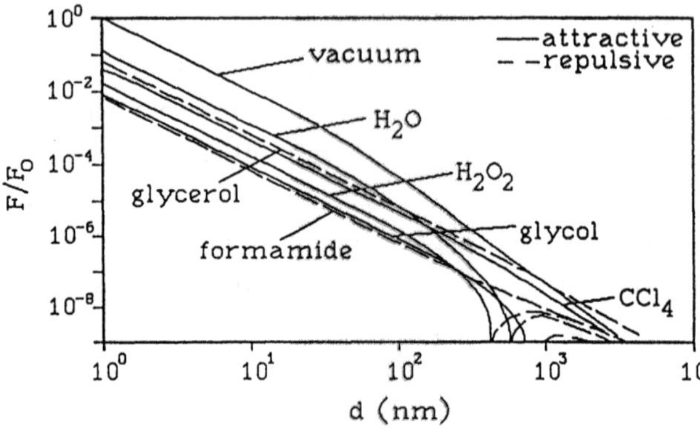

Fig.6.
vdW curves for a SiO_2-metal combination of probe and sample (all results of the theory are invariant to an exchange of probe and sample materials) immersed into various polar liquids. The probe is approximated by a paraboloid. Forces are normalized with respect to the vacuum level for d=1nm.

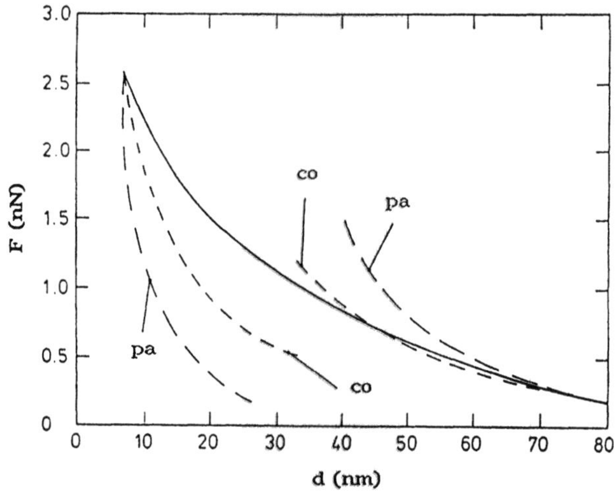

Fig.7.
Experimentally recorded vdW curve for a Ni tip and a HOPG sample (solid line) in comparison with theoretical data obtained for a paraboloidal (pa) and a conical (co) probe.

Figure 7 shows a vdW curve measured for a Ni tip and a HOPG sample. Up to probe-sample distances of about 40 nm forces are dominated by liquid contaminations on probe and sample [30]. However, at sufficiently large probe-sample distances, the experimental data clearly approach the theoretically expected curves, thus demonstrating the presence of retarded vdW interactions.

Experimental Setup

During the past few years several MFM devices have been presented [31]. We employ the capacitively controlled compact instrument [32] shown in Fig.8. The relatively simple device exhibits a force sensitivity in the 100pN range and is well suited for both, atomic resolution AFM and noncontacting modes of operation.

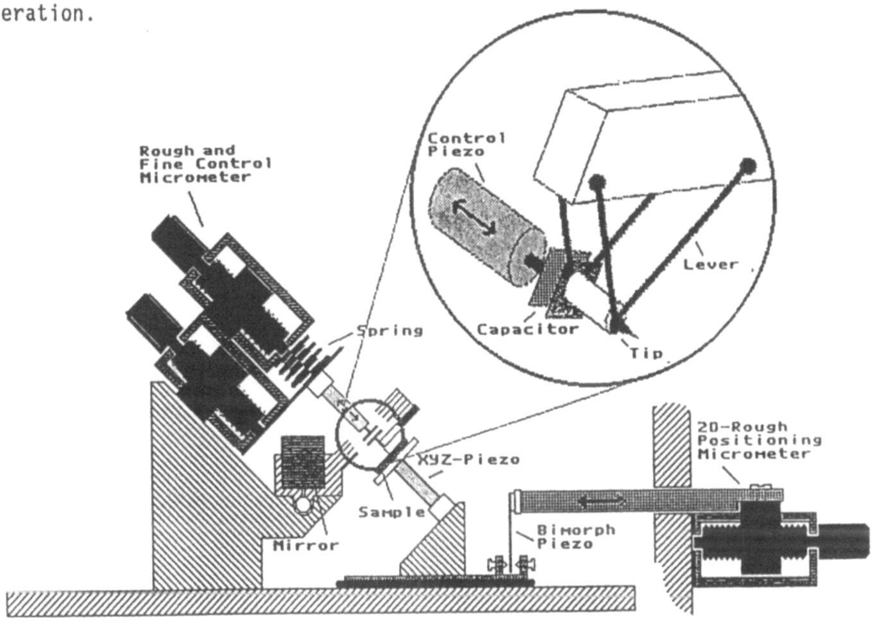

Fig.8.
Mechanical setup of the capacitively controlled force microscope.

Ferromagnetic microprobes are fabricated by improved electrochemical techniques, yielding sharp tips with apex radii down to 50nm [33]. Apart from these conventional probes, we use the new type of spherical microprobes [33] shown in Fig.9. The advantages of these microfabricated hybrid probes should be the straightforward alignment to the sample surface, a relatively simple internal magnetic configuration, and low self-fringe fields.

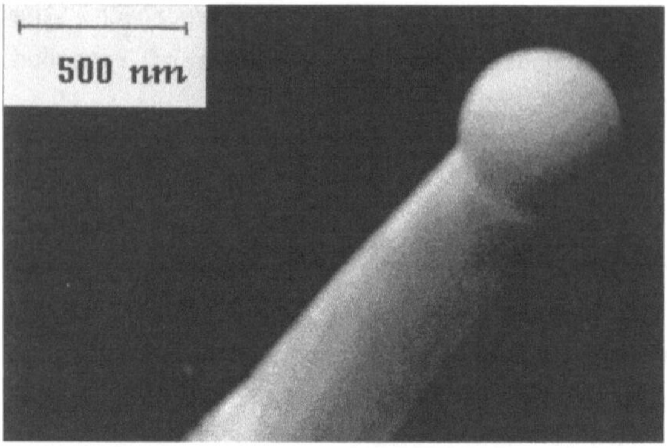

Fig.9.
Spherical hybrid microprobe. A Ni sphere is deposited at the very tip of a needle-shaped shank of Ni.

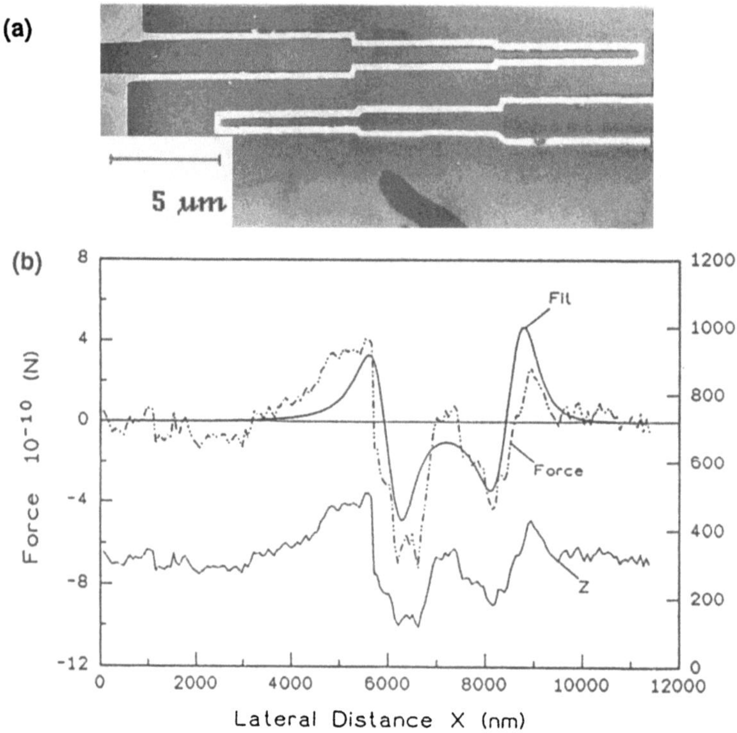

Fig.10.
(a) Nb double-wire planar test structures on an oxidized Si substrate with three different spacings between the wires: 0.5, 1.0, and 2.0μm. The cross section of a wire is 500nm x 50nm. (b) Data obtained for a single line scan at 2μm wire spacing. Variations of the tip sample-spacing (z) are detected in the constant-force mode. The magnetostatic force is reconstructed by accounting for the superimposed overall electrostatic servo force [Eq.(1)]. Fitting the theoretical data then yields the effective volume and internal magnetization for an employed probe model [34] (monopolar or dipolar probe, see Fig.2).

A New Method for Probe Calibration

As discussed above (Fig.2), the response function in MFM predominantly relies on the *a priori* unknown internal magnetic configuration of the probe. For experimental probe calibration the magnetic field created by lithographically formed, current-driven planar structures is well suited. While the microfield configuration of the test structure is exactly known, the effective detector volume and the internal magnetic configuration (monopolar or dipolar) of the probe are treated as free parameters to be fitted to the experimental data which are obtained for successive scans (at varying working distance) over the structure [34]. Figure 10 shows a typical result of a calibration experiment. Apart from providing important information about the MFM probes, the experiments show that MFM is well suited for imaging magnetic fields of highly integrated current-driven micropatterns; a new application of considerable technical relevance.

Domain Wall Imaging

In the following we present examples for domain wall analysis by MFM, demonstrating the considerable potential of the technique in basic micromagnetic research. The most striking feature of the MFM image in Fig.11, taken on high purity Fe [11], is the vanishing microfield of the central 180^0 Bloch wall in the left part of the image. The reason is presumably a subsurface extension of the domain arrangement, providing a nearly complete flux closure internal to the sample.

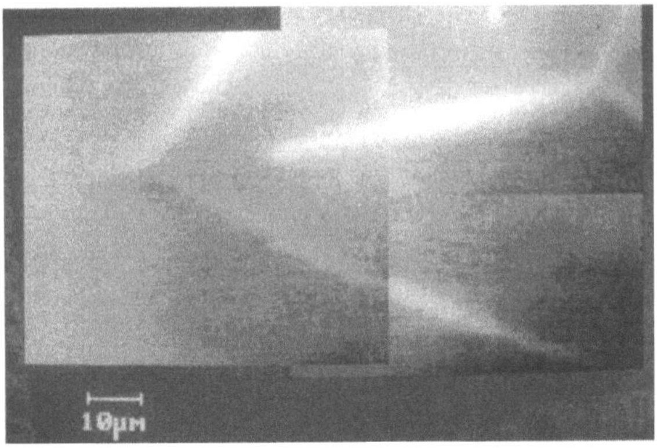

Fig.11.
Large-scale MFM image of a domain configuration in a Fe single crystal.

Figure 12 shows a high-resolution image of the intersection of two 90^0 walls with one 180^0 wall (as in the right part of Fig.11). Characteristic profiles of the individual walls as well as of the intersection area become visible and can be studied in detail by analyzing individual line scans [11,35-37].

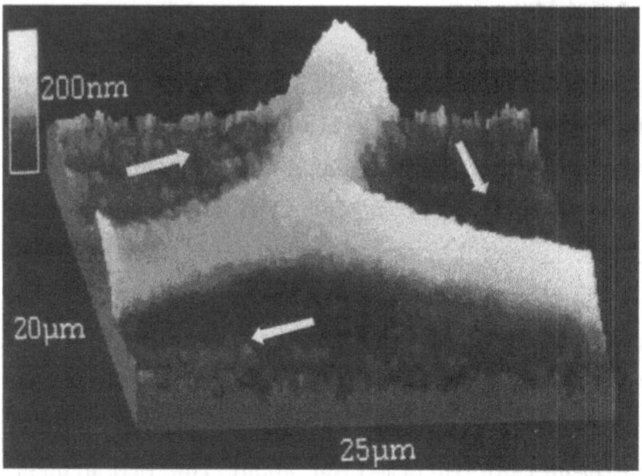

Fig.12.
MFM image of the intersection of two 90^0 Bloch walls with a 180^0 wall in Fe.

A transition from a spin-up to a spin-down orientation within a 180^0 Bloch wall is shown in Fig.13 [11,35]. Right at the transition between the adjacent segments, the wall contrast vanishes, indicating the existence of a Bloch line [38,39]. The latter configuration involves a micromagnetic singularity which is expected to strongly interact with crystalline lattice defects, thus having a significant effect on the magnetization reversal process of a ferromagnetic medium [40].

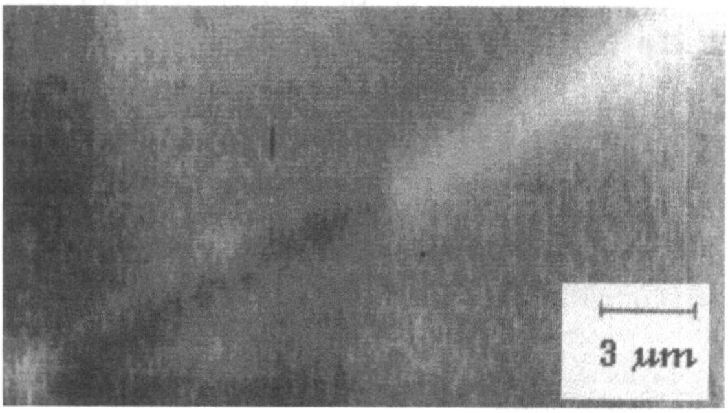

Fig.13.
Part of a bipolar 180^0 Bloch wall in Fe. The adjacent wall segments are separated by a Bloch line.

TUNNELING SPECTROSCOPY ON TYPE-II SUPERCONDUCTORS

Principles of Contrast Formation

STS relies on the spatially resolved mapping of the density of electronic surface states of a sample. The most convenient mode of operation is to periodically switch off the feedback loop of a STM and to record I(V) characteristics (or conductance curves) at any point of the sample surface. In this way occupied and empty electronic states are revealed. The quasiparticle density of states for a superconductor exhibits an energy gap of 2Δ, $\Delta=V_g e$, about the Fermi level, where no single-electron states exist. When the tunneling voltage between a normal conducting STM tip and the superconductor (SIN junction) exceeds V_g, then electrons see a sharp rise in the density of states and can tunnel between probe and sample. The spatial variation of Δ can thus be directly mapped by STS. Several attempts have been reported which were performed on Nb_3Sn [41], NbN [42], Nb [43], and on various high-Tc materials [44-48]; however, mainly due to surface quality problems only with moderate success.

If a type-II superconductor is subject to an appropriate external magnetic field, the flux penetrates the sample in terms of a hexagonal arrangement of cylindrical magnetic domains (left column of Fig.14). The vortices exhibit a normal conducting core region with vanishing energy gap within the superconducting environment. A straightforward method to sense the vortex lattice is to detect the tunneling current at a voltage V_0, smaller than the gap voltage V_g, while the feedback loop is switched off (middle column of Fig.14). The current then varies between I_1 (superconducting limit) and I_2 (normal conducting limit). When the feedback loop is turned on again, ordinary STM operation at a voltage V, large compared with V_g, gives the surface topography. Displaying the current $I(V_0)$ upon raster-scanning the surface then yields an image of the magnetic domain pattern (right column of Fig.14) [52,53].

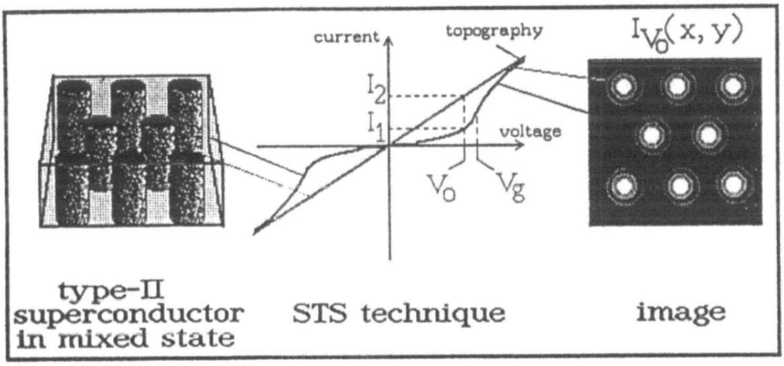

Fig.14.
Principle of contrast formation for the STS analysis of magnetic domains in type-II superconductors.

Experimental Setup

Low-temperature STS is performed with the compact device shown in Fig.15 which is operated in a bath cryostat, directly immersed into the liquid helium [53]. A superconducting coil produces homogeneous magnetic fields perpendicular to the sample plane. The STM is operated in the constant-current mode (5mV, 50pA), while spectroscopic information is obtained by intermediately interrupting the feedback loop and by alternatively detecting a complete I(V) characteristic or the current, corresponding to a fixed voltage V_0 within the gap width V_g of the superconducting sample [52,53] (see Fig.14).

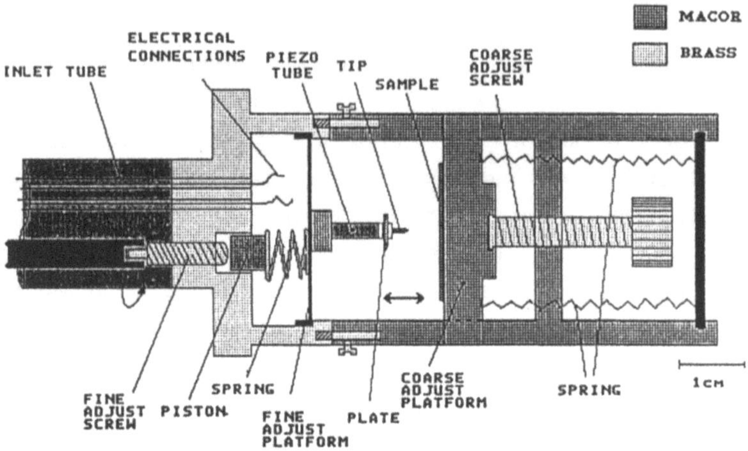

Fig.15.
Mechanical setup of the low-temperature STM.

The sample is single crystalline 2H-$NbSe_2$ [49], a layered type-II compount with a charge density wave (CDW) transition at ≈33K [50]. While a fraction of the Fermi surface is removed by the CDW gap [21], the remainder undergoes a BCS transition at 7.2K with ξ=7.7nm (in-plane) and λ=69nm [51]. The samples are cleaved immediately before installing them in the STM. Control observations generally show atomically flat surfaces extended over large areas [52,53]. As probes we use sharp tips of W, PtIr, and Ni.

Imaging of the Abrikosov Flux Lattice

Figure 16 shows a typical I(V) caharacteristic at zero applied field [52,53]. From numerous of these measurements we deduce a gap width (in-plane) of Δ=(0.9±0.1)meV at 4.2K. Since $2\Delta/kT$=3.2 is in reasonable agreement with BCS theory, we believe bulk rather than anomalous surface superconducting behavior is being probed. Thermal smoothing of the gap energy contributes to the considerable subgap current seen in Fig.16.

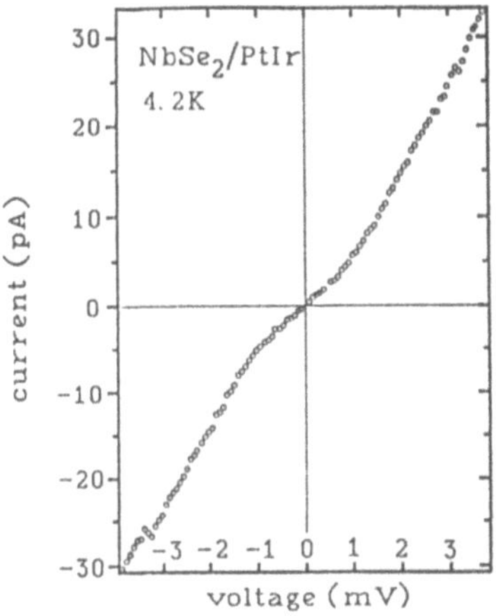

Fig.16.
Typical I(V) characteristic recorded at zero applied field.

Upon applying an external magnetic field, the sample is transferred to the superconducting mixed state and the Abrikosov flux lattice is formed as schematically shown in Fig.14. Figure 17 shows a STS image of the domain arrangement recorded with a PtIr tip [52,53]. At V_0=0.9mV≈V_g, a maximum variation of the current between normal conducting vortex core and superconducting environment of I_2-I_1=1.9pA is obtianed.

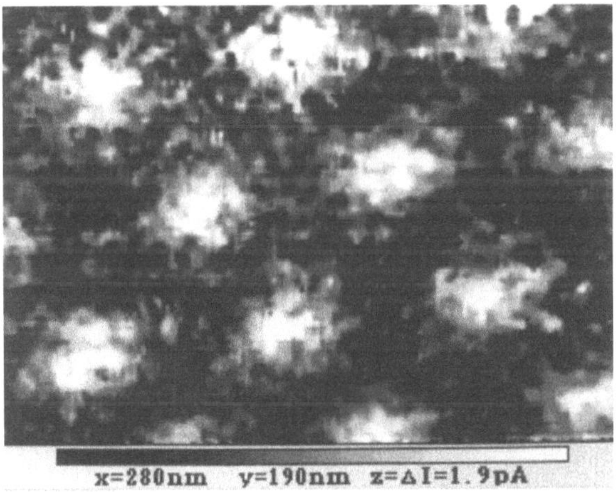

Fig.17.
Magnetic domains in $NbSe_2$ at 4.2K in a magnetic field of 0.34T perpendicular to the sample plane. The ohmic portion of the underlying I(V) characteristics (see Fig.14) corresponds to a tunneling resistance of 110MΩ. The image was recorded with a PtIr probe.

We repeated the above experiments, employing sharp ferromagnetic Ni tips. As stated above, these tips exhibit a strong magnetic stray field which is, for a conical probe model, about 1.3T (the upper critical field of $NbSe_2$ is expected to be 3.2T perpendicular to the sample surface [51]) at typical tunneling distances [see Fig.4: α=15°, z/l=0.001, μ_0M(Ni,4.2K)=0.64T]. The I(V) characteristics recorded at zero external field show no significant differences with respect to Fig.16 [52,53]. Figure 18 shows an Abrikosov flux lattice imaged with a Ni probe [52,53]. It is clearly obvious that the magnetostatic probe-sample interaction does not deform the vortex lattice in any notable way, in spite of the low internal pinning force of the $NbSe_2$ single crystals, for which we measured a critical current density of $J_c(0.5T,4.2K)=15A/cm^2$ [52,53].

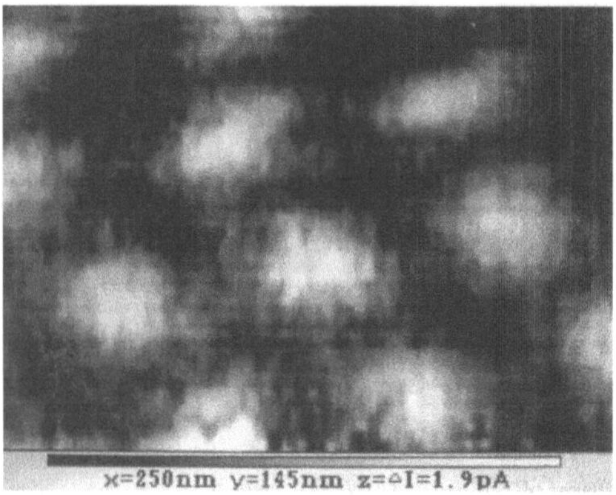

Fig.18.
Abrikosov vortex lattice imaged at 4.2K and 0.47T with a ferromagnetic tip. The tunneling resistance was about 100MΩ.

Concerning the technically more relevant superconductors, the low-temperature STS magnetic analysis is generally not possible since the sample surfaces are often highly corrugated and contaminated. The problem becomes clearly obvious for the high-T_c sample shown in Fig.19. The above results obtained by tunneling through ferromagnetic tips may open the possibility of using low-temperature MFM to image domains in type-II superconductors in a nondestructive way, avoiding surface quality problems by choosing a probe-sample spacing in the nm or even 10nm range.

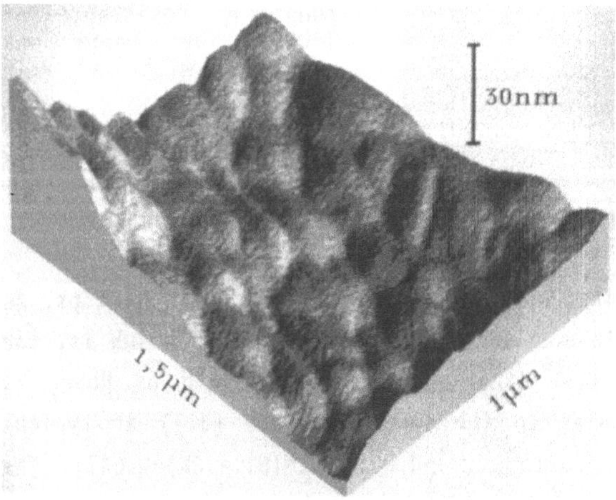

Fig.19.
AFM image of a sputter-deposited YBaCuO thin film. Surface corrugations and contaminations make STS magnetic analysis apparently difficult. Anomalous surface superconductivity is often being probed. [Sample courtesy of R. Wördenweber]

CONCLUSIONS

We presented an analysis of magnetic configurations in ferromagnets and type-II superconducting media performed by modern scanned probe techniques.

Imaging of domain boundaries by high-resolution MFM reveals micromagnetic details which are only hardly accessible to conventional observation techniques. The results are of current importance to basic micromagnetic research. The analysis of magnetic storage media, of thin film magnetic devices, and of highly integrated current-driven planar patterns is classified as a MFM application of considerable relevance to the microelectronic industry. However, to our opinion, the ultimate capabilities of the technique have not yet been obtained. The theoretical approach to contrast formation shows that predominantly improved microprobes are needed to optimize the MFM transfer function.

We succeded in imaging the Abrikosov vortex lattice in a type-II superconductor by employing low-temperature STS. Since tunneling through sharp ferromagnetic tips does not effect the superconducting behavior of the samples in any notable way, we propose low-temperature MFM as a high-resolution and nondestructive technique for imaging vortex configurations in technically relevant superconducting materials and devices, where surface quality problems often make STS difficult or even impossible.

1 *Martin Y, Wickramasinghe HK (1987) Appl Phys Lett 50:1455*
2 *Saenz JJ, Garcia N, Grütter P, Meyer E, Heinzelmann H, Wiesendanger R, Rosenthaler L, Hidber HR, Güntherodt HJ (1987) J Appl Phys 62:4293*
3 *Binnig G, Quate C, Gerber (1986) Phys Rev Lett 56:930*
4 *Grütter P, Meyer E, Heinzelmann H, Rosenthaler L, Hidber HR, Güntherodt HJ (1988) J Vac Sci Tecnol A 6:279*
5 *Grütter P, Jung T, Heinzelmann H, Wadas A, Meyer E, Hidber HR, Güntherodt HJ (1990) J Appl Phys 67:1437*
6 *Grütter P, Wadas A, Meyer E, Heinzelmann H, Hidber HR, Güntherodt HJ (1990) J Vac Sci Tecnol A 8:406*
7 *Grütter P, Wadas A, Meyer E, Hidber HR, Güntherodt HJ (1989) J Appl Phys 66:6001*
8 *Hobbs PCD, Abraham DW, Wickramasinghe HK (1989) Appl Phys Lett 55:2357*
9 *Mamin HJ, Rugar D, Stern JE, Fontana Jr RE, Kasiraj P (1989) Appl Phys Lett 55:318*
10 *Göddenhenrich T, Hartmann U, Anders M, Heiden C, (1988) J Microsc 152:527*
11 *Göddenhenrich T, Lemke H, Hartmann U, Heiden C, (1990) Appl Phys Lett 56:2578*
12 *Hartmann U (1990) J Magn Magn Mat 83:545*
13 *Mamin HJ, Rugar D, Stern JE, Terris BD, Lambert SE (1988) Appl Phys Lett 53:1563*
14 *Abraham DW, Williams CC, Wickramasinghe HK (1988) Appl Phys Lett 53:1446*
15 *Rugar D, Mamin HJ, Guethner P, Lambert SE, Stern JE, McFadyn I, Yogi T (1990) J Appl Phys 68:1169*
16 *Martin Y, Rugar D, Wickramasinghe HK (1988) Appl Phys Lett 52:244*

17 Schönenberger C, Alvarado SF (1990) J Appl Phys 67:7278
18 Binnig G, Rohrer H, Gerber C, Weibel E (1982) Phys Rev Lett 49:57
19 Hess HF, Robinson RB, Dynes RC, Valles Jr JM, Waszczak JV (1989) Phys Rev Lett 62:214
20 Hess HF, Robinson RB, Waszczak JV (1990) Phys Rev Lett 64:2711
21 Hess HF, Robinson RB, Dynes RC, Valles Jr JM, Waszczak JV (1990) J Vac Sci Tecnol A 8:450
22 Wiesendanger R, Güntherodt HJ, Güntherodt G, Gambino RJ, Ruf R (1990) Phys Rev Lett 65:247
23 Schönenberger C, Alvarado SF (1990) Z Phys B - Condensed Matter 80:373
24 Hartmann U (1989) phys stat sol (a) 115:285
25 Hartmann U (1989) Phys Lett A 137:475
26 Hartmann U (1990) J Vac Sci Technol A 8:411
27 Hartmann U (1988) J Appl Phys 64:1561
28 Hartmann U (1990) Phys Rev B 42:1541
29 Hartmann U (1990) Proc STM'90/NANO1 Conf, Baltimore, USA; J Vac Sci Technol (to be published)
30 Weisenhorn AL, Hansma PK, Albrecht TR, Quate CF (1989) Appl Phys Lett 54:2651
31 Martin Y, Abraham DW, Hobbs PCD, Wickramasinghe HK (1990) J Electrochemical Soc (to be published)
32 Göddenhenrich T, Lemke H, Hartmann U, Heiden C, (1990) J Vac Sci Technol A 8:383
33 Lemke H, Göddenhenrich T, Hartmann U, Heiden C, (1990) Rev Sci Instrum (October Issue)
34 Göddenhenrich T, Lemke H, Mück M, Hartmann U, Heiden C, (1990) Appl Phys Lett (to be published)
35 Hartmann U, Göddenhenrich T, Lemke H, Heiden C, (1990) Proc INTERMAG'90 Conf, Brighton, UK; IEEE Trans Magn (to be published)
36 Hartmann U, Heiden C (1988) J Microsc 152:281
37 Hartmann U (1989) Phys Rev B 40:7421
38 Hartmann U, Mende HH (1986) J Appl Phys 59:4123
39 Hartmann U, Mende HH (1986) Phys Rev B 34:4878
40 Hartmann U (1987) Europhys Lett 4:1079
41 deLozanne AL, Elrod SA, Quate CF (1985) Phys Rev Lett 54:2433
42 Kirtley JR, Raider SJ, Feenstra RM, Fein AP (1987) Appl Phys Lett 50:1607
43 Berthe R, Hartmann U, Heiden C (1988) J Microsc 152:831
44 Kirtley JR, Feenstra RM, Fein AP, Raider SJ, Gallagher WJ, Sandstro R, Dinger T, Schafer MW, Koch R, Laibowitz R, Bumble B (1988) J Vac Sci Technol A 6:259
45 deLozanne AL, Ng KW, Pan S, Silver RM, Berezin A (1988) J Microsc 152:117
46 Kirk MD, Eom CB, Oh B, Spielmann SR, Beasley MR, Kaputulnik A, Geballe TH, Quate CF (1988) Appl Phys Lett 52:2071
47 van de Leemput LEC, van Bentum PJM, Driesen FAJM, Gerritsen JW, van Kempen H, Schreurs JWM, Bannema P (1988) J Microsc 152:103
48 Gallgher MC, Adler JG (1990) J Vac Sci Technol A 8:464
49 The $NbSe_2$ samples were kindly supplied by F. Lévy, EPFL Lausanne
50 Coleman RV, Giambattista B, Hansma PK, Johnson A, McNairy WW, Slough CG (1988) Adv Phys 37:559
51 deTrey P, Gygax S, Jan JP (1973) J Low Temp Phys 11:421
52 Berthe R, Hartmann U, Heiden C (1990) Appl Phys Lett (to be published)
53 Berthe R, Hartmann U, Heiden C (1990) Proc STM'90/NANO1 Conf, Baltimore, USA; J Vac Sci Technol A (to be published)

Acoustic Microscopy: Pictures to Ponder

G.A.D. Briggs, R. Gundle[1], C.W. Lawrence, A. Rodriguez-Rey[2], and C.B. Scruby[3]

Department of Materials, University of Oxford, England OX1 3PH

Abstract

Applications of acoustic microscopy are growing in various areas of materials and biology. In metal, ceramic, and glass matrix composites the acoustic microscope enables you to look at the interface between the fibres and the matrix, the formation of cracks in the matrix, and phase transformations. In a particular case of Nicalon (SiC) fibres in a borosilicate glass matrix, a crystobalite phase was discovered in the matrix, due to devitrification during processing. Different phases have also been identified in rock samples. Biotite, plagioclase, and quartz can be distinguished in granodiorite. In another specimen of this rock, cracks, grain boundaries, and subgrain boundaries can all be seen in the same picture, and the relationships between them studied. In the medical world, acoustic microscopy is being used to study the development of the mechanical properties of matrix from bone-derived cells. While the cells are still living, the collaginized and mineralized fibrous structure and its elastic properties can be examined in detail.

Introduction

Acoustic microscopy enables you to image the interaction of acoustic waves with the elastic features of a specimen. An acoustic lens is used to focus a short wavelength acoustic wave to a diffraction limeted spot, and to detect the returning echo. The use of mechanical scanning in a microscope, which was something of a innovation when it was used for the acoustic microscope, is now familiar from many other kinds of scanning microscope too. It would be impossible to describe the full range of applications here, still less the instrumental details and the imaging theory, and so instead we shall describe some of our favourite pictures from our own work.

Composites

Quite a number of low-ductility materials such as ceramics and composites have been studied by acoustic microscopy [Scruby *et al.* 1989; Lawrence *et al.* 1989, 1990]. Figure 1(a) is an acoustic micrograph of a section through a glass matrix composite. The matrix is a borosilicate glass, similar to glass better known under the trade name Pyrex. The role of the matrix is largely to hold the fibres together (preferably without touching), and to protect their surfaces mechanically and chemically so that they can support loads at elevated temperatures in aggressive environments. The composition of the glass can be tailored for optimum compatibility with the fibres, and the relatively low viscosity at high temperature enables almost

---

[1]also at Nuffield Orthopaedic Centre, Oxford

[2]also at Department of Geology, University of Oviedo, 33080-Oviedo, Spain

[3]also at AEA Technology, Harwell Laboratory, Oxon. OX11 0RA

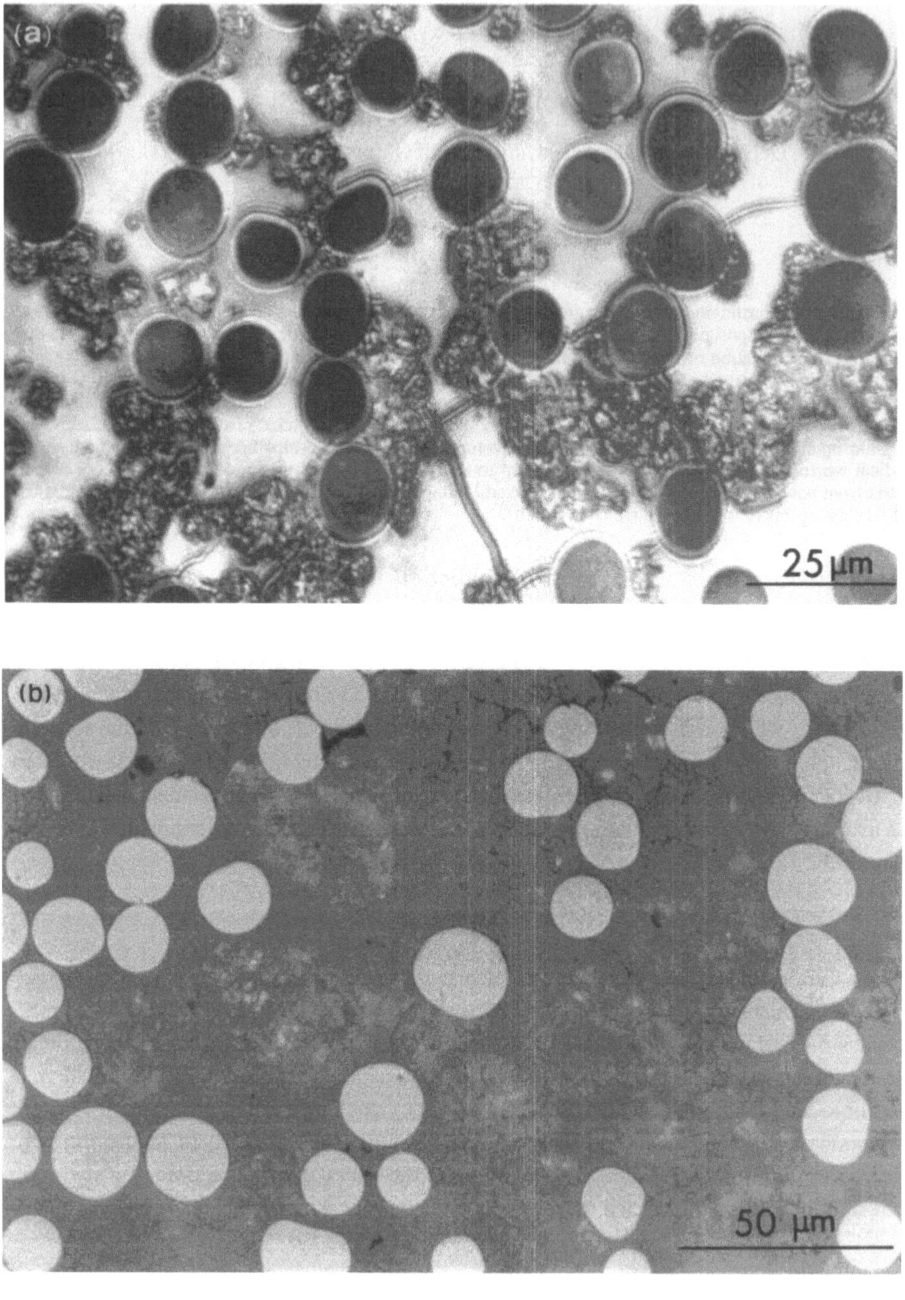

Figure 1. A section through a glass matrix composite: (a) acoustic micrograph, 1.9 GHz, $z = -2.8\ \mu m$; (b) optical micrograph. The matrix is borosilicate glass, Corning code 7740: composition by weight SiO_2, 81%; Al_2O_3, 2%; B_2O_3, 13%. The fibres are silicon carbide, manufactured by Nippon Carbon under the name Nicalon [Yajima *et al.* 1978]: nominal diameter, 15 µm; tensile strength, 2.75 GPa; Young modulus, 196 GPa; Poisson ratio, 0.15; density, 2.55 Mg m^{-3}. The composite has fibre volume fraction 0.49; flexural strength, 1.25 GPa; Young modulus, 120 GPa; fracture toughness, 70 MN $m^{-3/2}$.

complete penetration of the matrix into the fibre bundles during fabrication. The commercial name of the fibres is Nicalon, they are composed mainly of silicon carbide. The composite has a flexural strength of 1.25 GPa and a fracture toughness of 70 MN $m^{-3/2}$. For comparison, an optical picture of another area of the specimen is shown in figure 1(b). Newcomers to acoustic microscopy find optical comparisons reassuring, though as confidence is gained with the acoustic microscope such comparisons become less and less necessary. Crucial questions for the development of a composite like this one are: are its properties, such as its strength and toughness at high temperatures, the best that can ever be obtained? If so, how do you make sure that those properties are consistently achieved? If not, how do you improve them still further?

Great care must be taken not to introduce damage in the preparation of sections of ceramic composites for microscopy, otherwise it will be hard to distinguish what was intrinsic to the specimen from what was introduced during the preparation of the section. This specimen was first cut off perpendicular to the direction of the fibres using a slow annular diamond wheel (100-150 r.p.m.), with a flow of water to lubricate the cutting process and to keep it cool. The cut-off piece was mounted in a cold-cure epoxy resin, with bubbles removed to prevent them from becoming traps for diamond paste in subsequent polishing. When the resin had set, the edges were rounded off, and the specimen was mounted on a precision polishing jig. The surface was lapped and polished, with a relatively low load, using oil soluble 6 μm diamond paste of on a scrolled cast iron platten, and then in turn 1, 0.25 and 0.1 μm paste on a scrolled brass platten. The specimen was thoroughly cleaned in an ultrasonic bath between every stage. It was given a final polish with recirculating Syton-W30 (0.125 μm colloidal particles of silica suspended in water) with a thin coating of polyurethane foam on the platten. In the acoustic microscope you do not want any surface relief, and so this stage had to be finely judged, not too short or the damage from the 0.1 μm paste would not be removed, not too long or softer material would be polished away too much. The specimen was then examined in the acoustic microscope.

Figure 1(a) shows a wealth of detail. First, and most obvious, the fibres have a different contrast from the matrix. In the imaging conditions here, the fibres are darker than the matrix. By simply counting the fibres and measuring their diameters, the mean SiC fibre diameter is 16 μm, and the area fraction is 0.49. The matrix is composed of two phases. About 70% of the matrix area appears homogeneous, while the remaining 30% appears granular, with a grain size less than 10 μm. The bright, homogeneous region in figure 1(a) is borosilicate glass, behaving itself perfectly well. The crystalline phase was subsequently confirmed, using both energy dispersive and wavelength dispersive spectroscopy on an electron microprobe and also X-ray diffraction, to be crystobalite. Crystobalite is an allotrope of silica with a cubic crystal structure, which is stable at high temperature (1470-1713°C) under atmospheric pressure. The precipitation of crystobalite, known as devitrification, is likely to occur when borosilicate glass is held for a prolonged time between the glass softening temperature and the crystallization temperature. Unfortunately, the fabrication procedure for 7740/SiC composite involves exactly these conditions. An acoustic micrograph of a section through a monofilament specimen parallel to the fibre axis is shown in figure 2(a). The crystobalite seems to grow to form a layer 3-5 μm thick around the fibre, but it also seems to nucleate away from the fibre too.

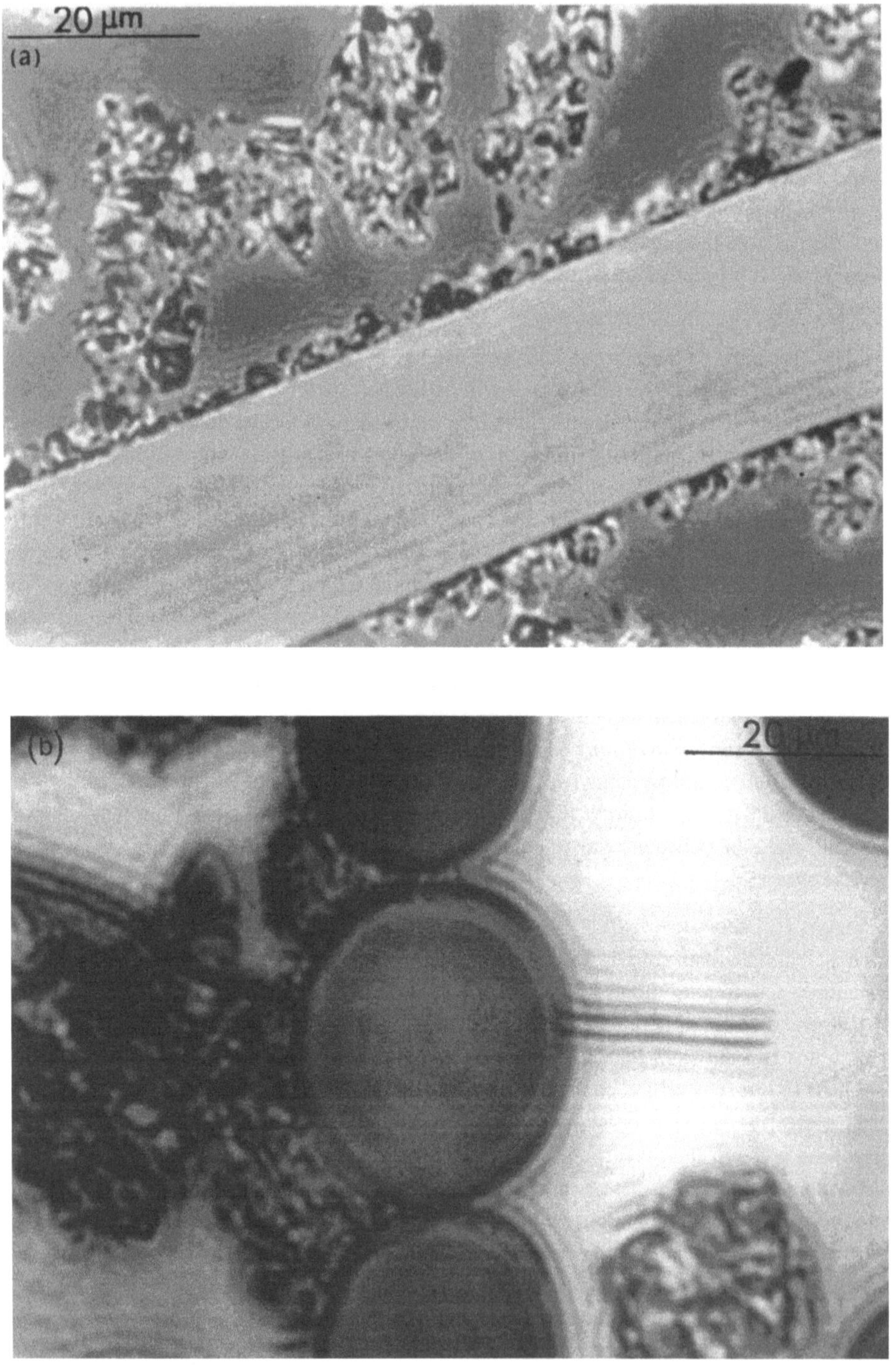

Figure 2. Acoustic images of the same 7740/SiC glass matrix composite as figure 1: (a) section parallel to the axis of a fibre; (b) section perpendicular to the fibre axes; 1.9 GHz.

There is a phenomenon in the acoustic images that is not seen by any other kind of microscopy. In all of the acoustic micrographs of these specimens there are fringes present. In figure 2(b) the contrast has been optimised to show them, both in the glass and in the silicon carbide. In the silicon carbide they are more widely spaced than in the glass, and they could be mistaken for growth rings, perhaps associated with the processing of the fibres, such as are found in a tree trunk. But they are not growth rings, they are an interference effect associated with waves that can be excited in the surface of solids. These waves are called Rayleigh waves after the great Cambridge scientist Lord Rayleigh, who discovered them theoretically in 1885. The spacing of the interference fringes is half the Rayleigh wavelength. The Rayleigh wavelength in the fibres must be greater than it is in the matrix, because the separation of the fringes is smaller in the matrix than it is in the fibres. In the matrix the fringes are associated with at least three kinds of features. First, there are fringes around the fibres, that are loci of constant distance from the interface between the fibre and the matrix, just as the fringes inside the fibres are. Then there are fringes, too, that seem to run around areas of crystobalite, or perhaps at constant distance from interfaces between crystobalite and glass. Finally, especially in figure 1(c), there are fringes that seem to run alongside cracks in the glass. The acoustic microscope has a greatly enhanced sensitivity to cracks or boundaries that run up to the surface, and is able to give strong contrast from them even when they are much finer than the wavelength being used for imaging. This sensitivity is associated with the excitation of Rayleigh waves in the surface, and their scattering by cracks or change of velocity as they pass from one material to another. The fringes are a manifestation of Rayleigh wave activity.

Rocks

A pair of pictures of an area of a polished 30 μm thick specimen of granodiorite, a type of granitic rock, are shown in figure 3. Once again great care was taken to obtain a flat surface with no relief polishing, though in this case it was possible to use standard techniques for preparation of such specimens for conventional petrographic study (except that only one surface is needed for acoustic microscopy in reflection). Three different minerals are present in this area of the specimen. In the top right corner is biotite, a highly anisotropic phyllosilicate. In the top left region is plagioclase, which appears in the rock as zoned crystals ranging in composition from oligoclase to andesine. Finally, the mineral in the lower region of the picture is quartz.

The difference between the pictures is due to the different distance from the lens at which each was taken. No-one can operate an acoustic microscope for very long without discovering that the contrast can vary dramatically as you move the specimen up and down relative to the lens. Indeed, it is quite possible for contrast reversals to occur between, for example, one grain in a specimen and another. In figure 3 the contrast of the biotite varies quite markedly between (a) and (b), relative to the plagioclase and the quartz. Plagioclase and quartz give rather similar contrast to each other in both pictures; they are distinguished by the twinning microstructure in the plagioclase.

In the disposal of radioactive waste, one potential pathway of radionuclides back to the environment is through cracks in the rock in which it is stored. Of course, this would only be relevant if a container failed and released waste into the surrounding rock, but it is important to understand what would happen in that case. Figure 4 is a series of images of a section of

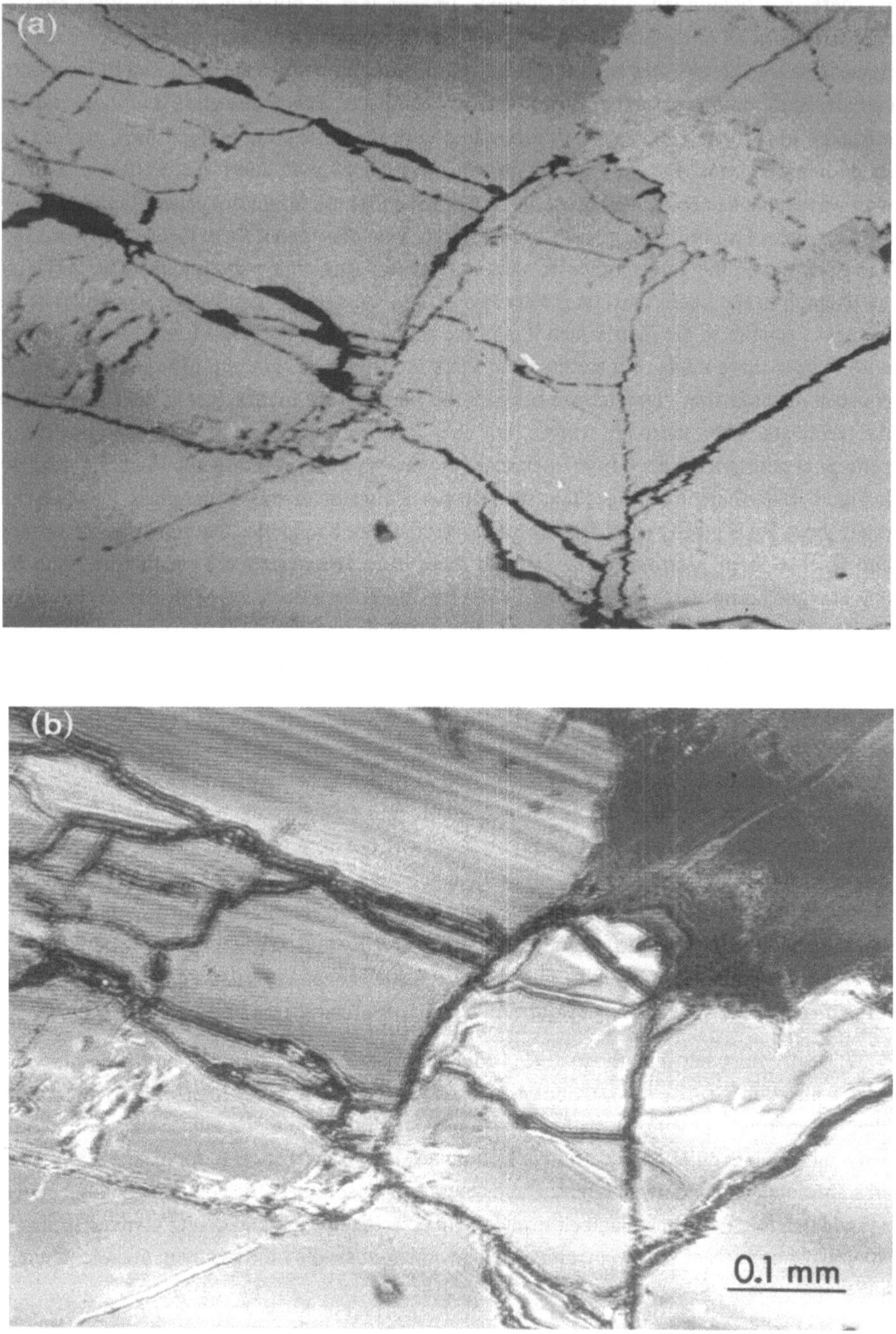

Figure 3. Polished granodiorite section containing plagioclase, biotite, and quartz; 400 MHz: (a) $z = 0$ μm; (b) $z = -10$ μm. (Rodriguez-Rey *et al.* 1990)

granodiorite that might be a candidate type of rock for the disposal of nuclear waste. Figure 4(a) is an acoustic image. The specimen had previously been studied by light microscopy and electron microscopy. By impregnating it with fluoresceine, and then examining it in a fluorescence microscope, it is possible to make the cracks show up as bright against an otherwise dark background; this is shown in figure 4(b). In figure 4(a) the same cracks show up with a characteristic fringe pattern like the ones in the acoustic pictures of the composite. Comparison of figure 1(a) and (b) suggests that both techniques are rather effective for revealing cracks. Rock in sections 30 μm thick can be imaged in a transmitted light polarizing microscope; figure 4(c) illustrates how the grain structure can be revealed in this way, because of the crystal optical anisotropy. On the left side of the picture there is a grain that looks uniformly bright, and which can be interpreted as a single quartz grain with only a few cracks in it. The other half of the picture has a grain with a different brightness (and a different colour, not reproduced here), which has rather more detail in it. If you look very closely, it is possible to discern rather fuzzy borders delineating different levels of brightness within that grain. But it is not easy. These borders correspond to the two slightly wiggly lines in the upper right quadrant of figure 4(a). They are boundaries where small misorientation within a grain has occured by the formation of dislocations in order to accommodate strain during the geological processes of deformation of the rock. It is blurred in the transmission polarized light micrograph because the picture is averaged over the whole 30 μm thickness of the sample. The contrast in the acoustic microscope comes from the slight change in crystallographic orientation across the misorientation boundary.

There is a particularly interesting crack indicated by * in the acoustic micrograph of this specimen. It is detected and identified as a crack in the acoustic microscope by the characteristic fringes that run either side of it. This crack had not been detected in any other microscope before it was seen in the acoustic microscope. The specimen was subsequently examined in a scanning electron microscope. By working first at low magnification, in order to identify the area corresponding to figure 4(a), and then progressively increasing the magnification, it was possible to find the crack. It occupies most of the diagonal in the s.e.m. picture in figure 4(d). Even so, it is not easy to see, and the s.e.m. picture emphasises the astonishing enhancement of the contrast from surface cracks that is available in the acoustic microscope.

Biological matrix

For a long time there has been a desire to understand better the relationship between the biomechanics and the biochemistry of bone development. The scientific motive is to understand how the mechanical properties develop on a microscopic scale as a function of time, and also in response to chemical and mechanical stimuli. The medical motive is to understand the interaction with prosthetic materials and devices, with a view to developing better materials and procedures for operations such as joint replacements. Hip and knee replacements for people over the retiring age have for some time been two of the most successful operations in the more developed world; it would be very nice to be able to extend them to younger age groups, and to other joints [Bulstrode 1987]. Surgeons at the Nuffield Orthopaedic Centre in Oxford are using acoustic microscopy to study the development of the mechanical properties of the matrix secreted by bone-derived cells. This is the material that eventually achieves the mechanical

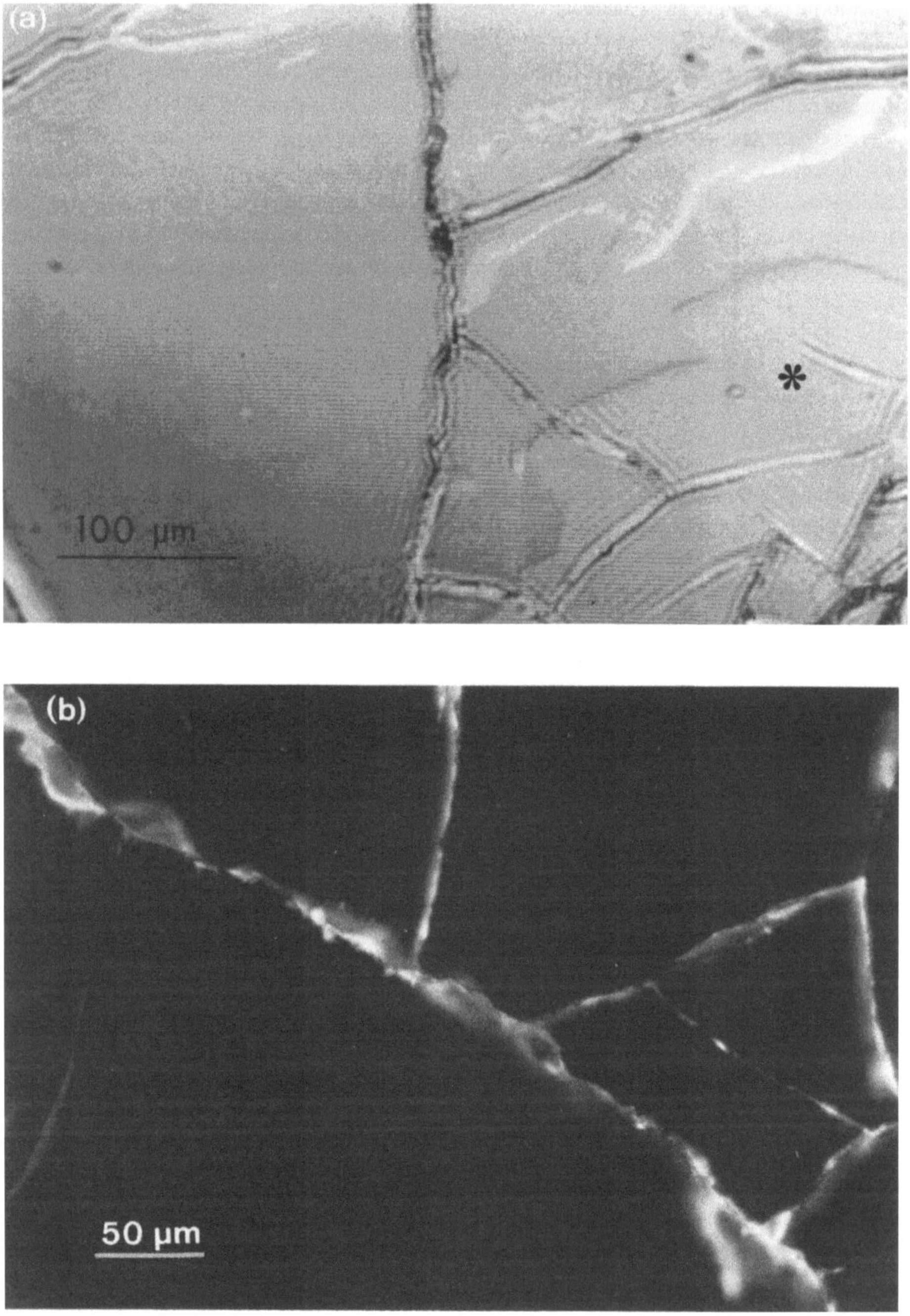

Figure 4. Polished granodiorite section: (a) 350 MHz, z = -10 μm; (b) reflected fluorescence microscopy *(continued on next page)*.

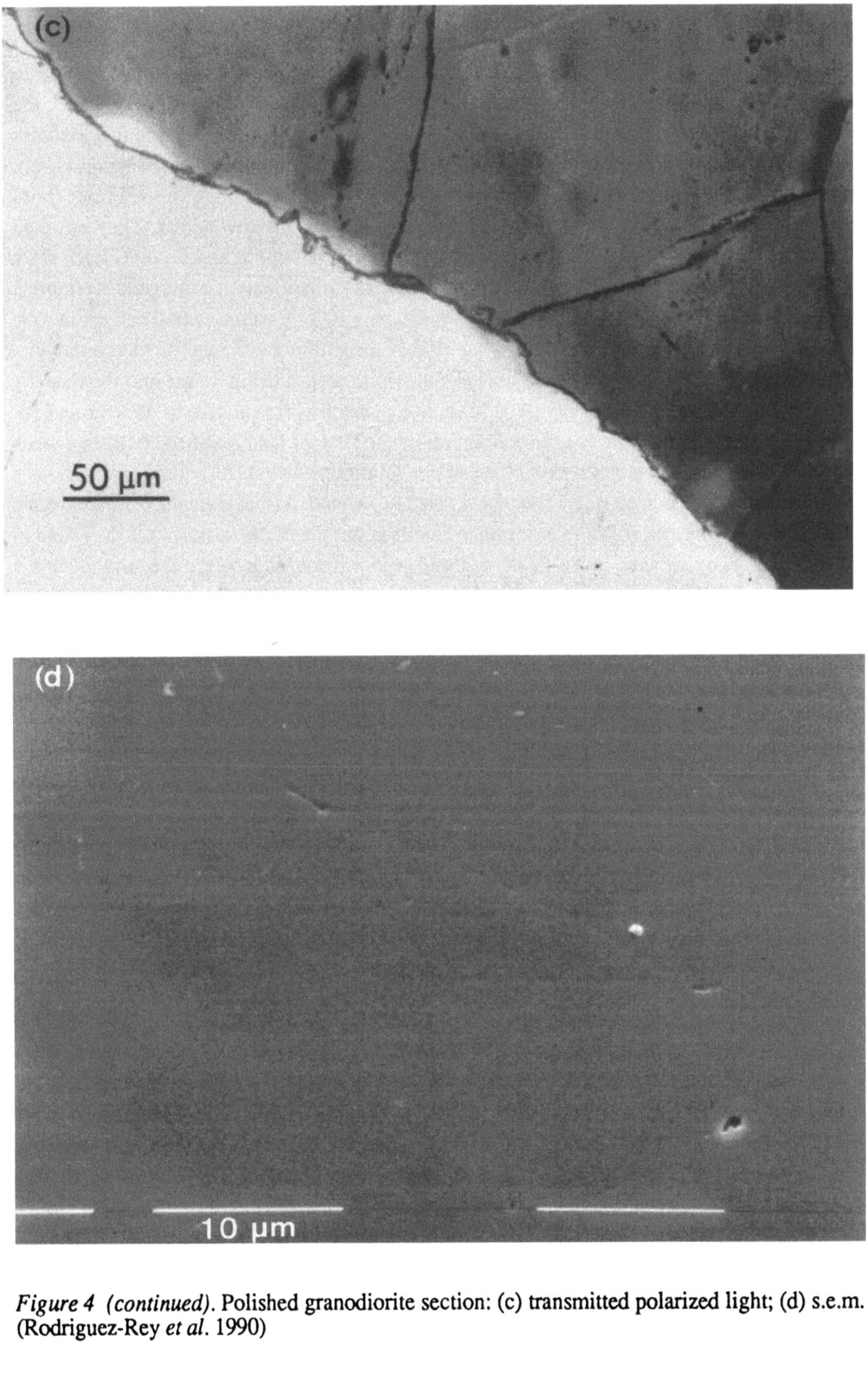

Figure 4 (continued). Polished granodiorite section: (c) transmitted polarized light; (d) s.e.m. (Rodriguez-Rey *et al.* 1990)

properties associated with bone. Pictures of two specimens, taken by Roger Gundle, are shown in figure 5.

The preparation of human bone-derived cells for such a study is a major undertaking in itself. The details are somewhat technical but very important [Beresford *et al.* 1983a,b]; the cells are described as bone-derived because they are a heterogeneous population that includes osteoblasts. During orthopaedic surgery, pieces of cancellous bone, the spongy part, are obtained sterile so that they are free from micro-organisms. The pieces are washed three times in phosphate-buffered sodium chloride to remove blood and marrow debris. They are then placed in special Nunc tissue culture flasks that have a flat bottom of area 25 cm^2. Each flask contains 10 ml of culture fluid, consisting of Dulbecco's modification of Eagle's medium, a mixture of nutrients including salts and amino acids, plus 10% by volume foetal calf serum (the fluid part of blood) supplemented with 50 units ml^{-1} penicillin and 15 $\mu g\ ml^{-1}$ streptomycin. The purpose of the antibiotics is to kill off any bacterial growth that may get into the flask by mistake. The bone explants are kept at 37°C in an incubator with a mixture of 95% air and 5% carbon dioxide. The medium is changed once after the first 24 hours, and three times a week after that. The cells start to grow out of the explant within seven days. After three to four weeks the whole area of the bottom of the flask is completely covered. At that point the cells are treated with trypsin, an enzyme which releases them from the surface of the flask so that they float in the medium. They are then given fresh medium, separated out in a centrifuge, and given yet more fresh medium. A sample of the cells is treated with trypan blue dye, which distinguishes between dead and living cells in a light microscope, so that you can see what proportion have survived all this tender loving care, and another sample is counted in a haemocytometer. 25,000 cells are then placed onto a 30 mm Falcon tissue culture dish, which has surface charge characteristics appropriate for bone-derived cells. The cells adhere well to the surface of the material that these dishes are made of, and the dishes fit directly on the stage of an acoustic microscope. The tissue culture dish is put inside a standard 10 cm Petri dish to keep out further unwanted micro-organisms, and the whole is placed in the incubator as before. The medium continues to be changed three times a week, and you can see how the cells are getting on by looking at them in a phase contrast microscope with green light. The rate of growth of the cells, and especially how much matrix they produce, how fast they produce it, and the rate at which it mineralises, can be controlled by varying the concentration of foetal calf serum in the medium, and also the concentrations of additives such as ascorbate, beta-glycerolphosphate and dexamethasone.

The micrograph in figure 5(a) came from a 5-week-old preparation. It was fixed in alcohol, and stained for alkaline phosphatase and, with von Kossa stain, for biomineral material. The biomineral material of interest here is hydroxyapatite, the principal crystalline mineral constituent of bone. The ordered structure visible within the matrix is not seen with either the light or electron microscopes. But the acoustic microscope can also work perfectly well with unfixed, unstained specimens. Figure 5(b) is an acoustic micrograph of matrix and cells from a 17-year-old male. In addition to the standard ingredients of culture medium, the cells were specifically stimulated with beta-glycerolphosphate and a vitamin C preparation. Because the acoustic microscope can look at the cells live, the same culture could be examined again and again over an extended period of time. Figure 5(b) came from an examination in the acoustic microscope after one week. One of the difficulties with these cells was that they wanted to

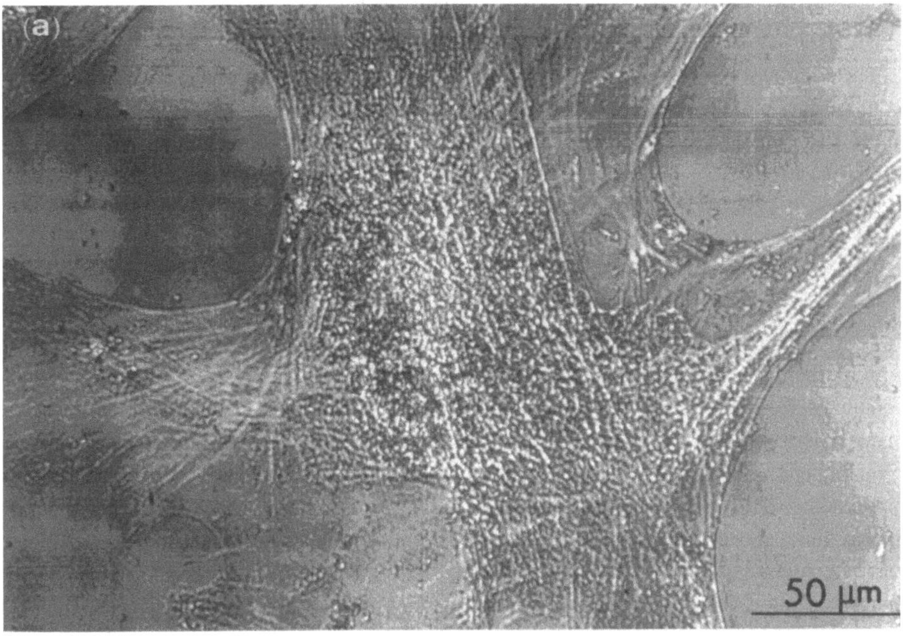

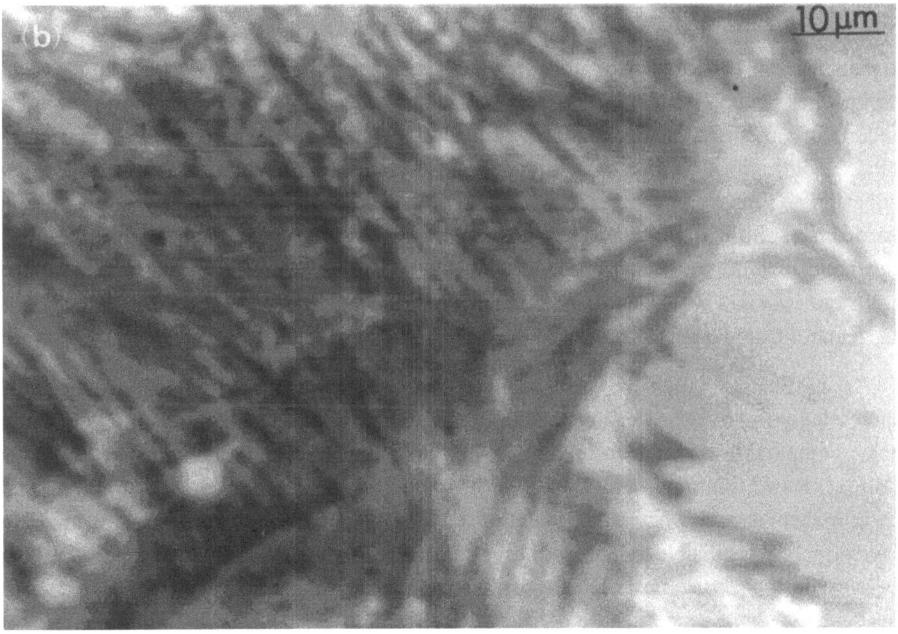

Figure 5. Matrix from human bone-derived cells: (a) fixed, after 5 weeks, 1.5 GHz; (b) unfixed, after 1 week, 1.3 GHz; both pictures were taken with the coupling fluid at 37°C and with the substrate at the focus of the lens.

climb on top of one another as they grew. Nevertheless, there is still a fascinating amount of detail about the matrix structure.

The significance of the acoustic micrograph here does not come solely, or even primarily, from the fact that the cells are living. It comes also from the fact that what you are seeing in figure 5 is contrast that comes from the mechanical structure of the specimen. In cases where this is the property of primary interest, acoustic microscopy offers the possibility seeing it directly with submicron resolution.

What else?

Applications of acoustic microscopy are growing so fast that it is impossible to keep up with them all, still less to include them all in a book like this. The references in this book contain no less than thirty three important publications from 1989-90, and no doubt there are many others too. Reviews over the last decade have included applications to human retina, revealing epithelium, red blood cells, collagen and melanin, Chinese hamster ovary cells, steel, integrated circuits and the interior of opaque objects [Wickramasinghe 1983]; printed circuit boards and a section of a cat's ear [Attal 1983]; grain structure in a transparent ferroelectric ceramic, quartzite, a eutectic CaF_2-MgF_2 synthetic crystal, defects in a $AgGaSe_2$ single crystal and in an epi-silicon layer, glass fibre reinforced polycarbonate, triballoy powder metal, a delaminated permalloy film on crown glass, photoresist inclusions in an integrated circuit and pinhole defects in the metalization of another integrated circuit, and endothelial cells from Xenopus laevis tadpole hearts [Hoppe and Bereiter-Hahn 1985]; alumina, silicon carbide, sialon, reaction-bonded silicon nitride, hot-pressed silicon nitride, zirconia, and ceramic-matrix composites [Fatkin *et al.* 1989]; granodiorite, caries lesions in human teeth, a histological section of a mouse muscle, creep crack growth in cermets, non-destructive testing of silicon nitride ball bearings, the fibre-matrix interface in a ceramic fibre composite [Briggs *et al.* 1989]; adhesion and disbonding, surface layers on GaAs and silicon, cracks in cermet coatings on hardmetals, and partially stabilised zirconia [Briggs 1989]; and semiconductor packaging, magnetic recording media, paint, and photo resist [Briggs and Hoppe 1990]. Quantitative techniques are now well developed too, so that it is possible to measure the elastic properties of the features that you see in an acoustic microscope.

As applications such as these become established, quantitative techniques become increasingly important, to enable the elastic properties of the specimen to be measured on a scale comparable with the resolution of the microscope. Several techniques are available [Briggs, Rowe, Sinton and Spencer 1989]. At first, the user had to choose between either measurement accuracy or spatial resolution, but he could not have both at the same time. But increasingly it is possible to have your cake and eat it. The line-focus-beam technique has long been established as the most accurate quantitative method for specimens, including anisotropic specimens, whose properties do not vary significantly over a region of a millimetre or so [Kushibiki and Chubachi 1985]. But now micro-line-focus-beam lenses are available that can measure Rayleigh wave velocity and attenuation in an anisotropic surface with a spatial resolution of 6 μm [Kushibiki, Chubachi and Tejima 1989]. The ability to measure surface stress has many important applications; surface stress has now been measured in a 15 μm thick alumina film as a function of direction with a resolution of 5-10 μm and a sensitivity of 0.3 MPa [Meeks *et al.* 1989]. Thin polymer coatings and sections of biological tissue can be measured by time-resolved techniques in which the echoes from the top and bottom interfaces are measured separately [Sinton *et al.* 1989; Daft and Briggs 1989], and this can now be used for areas of living cells that are as thin as 2 μm, with comparable spatial resolution [Wang *et al.* 1990].

Acknowledgements

This contribution is adapted from Chapter 1 of *Acoustic Microscopy*, to be published by Oxford University Press in 1991. We are grateful to OUP for allowing us to use this material here. Many colleagues have collaborated in the work described here: we especially wish to thank Dr B. Derby, Professor M. Montoto, Miss J. Wang, Dr C. Bulstrode, Dr M. Francis, and Mr K. Krämer. The use of the ELSAM at Oxford is in collaboration with Leica, Wetzlar.

References

Attal, J. (1983). La microscopie acoustique. Recherche 14, 664-7.

Beresford, J.N., Gallagher, J.A., Gowen, M. et al. (1983a) Human bone cells in culture. A novel system for the investigation of bone cell metabolism. Clin. Sci. 64, 38-39.

Beresford, J.N., MacDonald, B.R., Gowen, M. *et al.* (1983b). Further characterisation of a system for the culture of human bone cells. Calcif. Tiss. Int. 35, 637A.

Briggs, G.A.D. (1989). How sensitive is acoustic microscopy? EUROMAT '89 Proceedings, European Conference on Advanced Materials and Processes.

Briggs, G.A.D., Daft, C.M.W., Fagan, A.F., Field T.A., Lawrence, C.W, Montoto, M., Peck, S.D., Rodriguez-Rey, A. and Scruby, C.B. (1989). Acoustic microscopy of old and new materials. *In* Shimizu, H., Chubachi, N., Kushibiki, J. (Eds), Acoustical Imaging 17, 1-16. Plenum Press: New York.

Briggs, G.A.D., Rowe, J.M., Sinton, A.M. and Spencer, D.S. (1988). Quantitative methods in acoustic microscopy. IEEE 1988 Ultrasonics Symposium, 743-749.

Bulstrode, C. (1987) Keeping up with orthopaedic epidemics. Br. Med. J. 295, 514.

Daft, C.M.W. and Briggs, G.A.D. (1989a). Wideband acoustic microscopy of tissue. IEEE Trans UFFC 36, 258-263.

Fatkin, D.G.P, Scruby, C.B. and Briggs, G.A.D. (1989). Acoustic microscopy of low ductility materials. J. Mater. Sci. 24, 23-40.

Hoppe, M. and Bereiter-Hahn, J. (1985). Applications of scanning acoustic microscopy. IEEE Trans SU-32, 289-301.

Kushibiki, J. and Chubachi, N. (1985). Material characterization by line-focus-beam acoustic microscope. IEEE Trans SU-32, 189-212.

Kushibiki, J., Chubachi, N. and Tejima, E. (1989). Quantitative evaluation of materials by directional acoustic microscope. Ultrasonics International 89, 736-743.

Lawrence, C.H., Scruby, C.B. and Briggs, G.A.D. (1989). A study of ceramic fibre composites by acoustic microscopy. Inst. Phys. Conf. Ser. 98, 139-142.

Lawrence, C.W., Scruby, C.B., Briggs, G.A.D. and Dunhill, A. (1990). Crack detection in silicon nitride by acoustic microscopy. NDT International 23, 3-10.

Meeks, S.W., Peter, D., Horne, D., Young, K. and Novotny, V. (1989). Microscopic imaging of residual stress using a scanning phase-measuring acoustic microscope. Appl. Phys. Lett. 55, 1835-7. Cf. Residual stress mapping with a scanning phase-measuring microscope, IEEE 1989 Ultrasonics symposium, 809-812.

Rodriguez-Rey, A., Briggs, G.A.D., Field, T.A. and Montoto, M. (1990). Acoustic microscopy of rocks. J. Microsc. 160 (in the press).

Scruby, C.B., Lawrence, C.W., Fatkin, D.G.P., Briggs, G.A.D., Dunhill, A., Gee, A.E. and Chao, C.-L. (1989). Non-destructive testing of ceramics by acoustic microscopy. Br. Ceram. Trans. J. 88, 127-132.

Sinton, A.M., Briggs, G.A.D., and Tsukahara, Y. (1989). Time-resolved acoustic microscopy of polymer coatings.In Shimizu, H., Chubachi, N., Kushibiki, J. (Eds), Acoustical Imaging 17, 87-94.

Wang, J., Gundle, R. and Briggs, G.A.D. (1990). The measurement of acoustic properties of living human cells. Trans R. Microsc. Soc. (in the press).

Wickramasinghe, H.K. (1983). Scanning acoustic microscopy: a review. J. Microsc. 129, 63-73.

Yajima, S., Hasegawa, Y., Hayashi, J. and Iimura, M. (1978). Synthesis of continuous silicon carbide fibre with high tensile strength and high Young's modulus. J. Mater. Sci. 13,2569-76.

REAL-TIME CONFOCAL SCANNING MICROSCOPE - AN OPTICAL INSTRUMENT WITH A BETTER DEPTH RESOLUTION

T. Sure

Leica Mikroskopie und Systeme GmbH
Ernst Leitz Straße
D-6330 Wetzlar

ABSTRACT

In a confocal optical microscope the defocused image disappears instead of blurring, and therefore it is possible to examine different depth sections of a sample. Due to the increased contrast ratio this type of microscope is able to resolve two points whose distance is 30% closer than that for a normal microscope. The basic features for the image formation of such microscopes is given, and a special design which is able to produce the image in real-time will be discribed.

I. INTRODUCTION

The Confocal Scanning Optical Microscope (CSOM) was invented over 20 years ago [1, 2, 3]. In the last few years, however, it has become an important new imaging and measuring tool, especially for semiconductor materials. In the CSOM the image intensity drops off as the sample is defocused. So, in the confocal scanning microscope the defocused image disappears instead of blurring as in standard microscopes. This property makes it possible to image different depth layers of the sample. Thus, a three-dimensional reconstruction of the surface profile structure of the sample can be done with the help of electronic imaging systems. Furthermore, CSOM images have sharper edges and more contrast than those obtained by using standard microscopes. This leads to a better lateral resolution. To get a complete image of the sample it is necessary to scan either the sample or the light source. A special type of CSOM is the Real-Time Scanning Optical Microscope (RSOM) [4], which allows the operator to view different depth sections in real-time through the eyepieces without need of electronic recording. The difference between a CSOM and a RSOM will be described below.

II. BASIC PRINCIPLE OF THE CSOM

A conventional microscope consists of an illumination source, an objective lens and an eyepiece. In general, the sample is illuminated homogeneously, a technique known as the Köhler illumination. The objective lens forms a real, inverted image of the object in the intermediate image plane. The eyepieces produce a magnified, virtual image of the object. The total magnification of the system is the product of the objective and eyepiece magnifications. It is possible to achieve values in the range of 1000x - 2000x.

The CSOM differs from a conventional microscope by inserting one additional component, a small pinhole. This is shown schematically in figure 1. The illumination light passes through the pinhole and is focussed by the objective lens to a diffraction limited spot on the object. Light reflected from the sample passes the objective and goes back through the pinhole. If the object is moved out of focus, the reflected light is defocused and most of the light can not reach the detector. So, the amplitude of the detector signal drops rapidly as the light is reflected from surfaces which do not lie in the focal plane. This principle makes the CSOM suitable for the analysis of different depth layers of a sample.

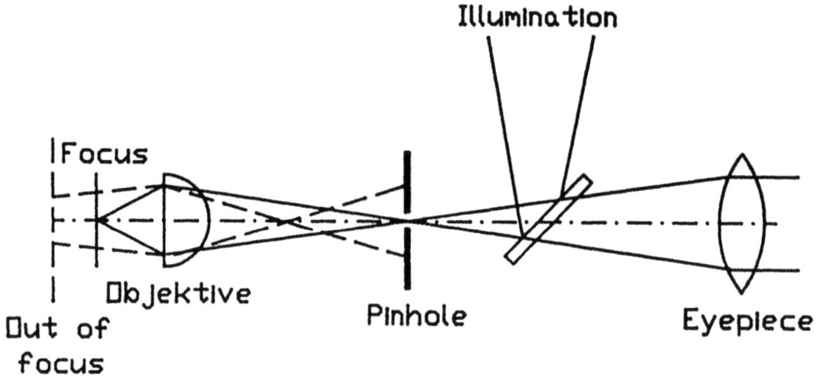

Fig.1: Principle path of ray for a CSOM. The dashed line shows the optical path for a plane which lies not in the focal plane.

The microscope is confocal because the objective lens is used twice, for illumination and for imaging. There are three basic features of a CSOM: point illumination, point detection and confocal imaging. There are some special designs to assemble a confocal optical microscope, in especially the use of a rotating Nipkow disk instead of a single pinhole. This type of microscope is able to produce a real-time image of the sample. Before describing the Real-Time Scanning Optical Microscope I consider it as useful to have a look at the theoretical background of image formation first and then to discuss the resolution of a confocal scanning microscope.

III. IMAGE FORMATION

This chapter contains a short survey of the formulas which are needed to discuss the image formation in a microscope. For more detailed information see for example [5, 6.].

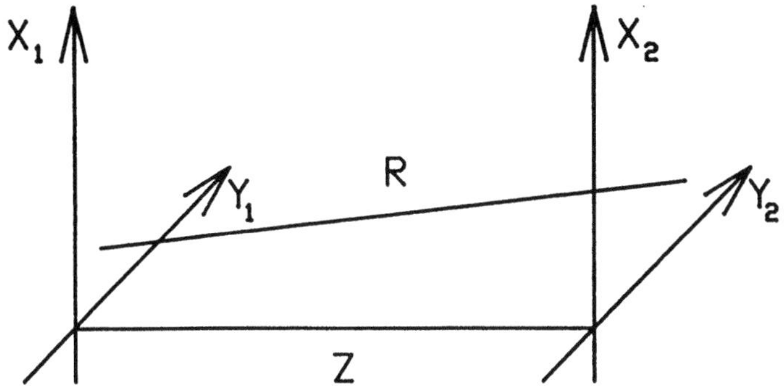

Fig.2: The diffraction geometry.

You can discribe the image in a plane (x_2,y_2) of an illuminated point in a plane (x_1,y_1) which are seperated by a distance R (see fig. 2) with the help of the Kirchoff diffraction formula:

$$U_2(x_2,y_2) = \int\int_{-\infty}^{\infty} \frac{1}{i\lambda R} U_1(x_1,y_1) \cdot \exp(-ikR)dx_1dy_1 \qquad (1)$$

We get the Fresnel approximation if we replace the R in the denominator by the distance z and expand the R in the exponent by the binominal theorem.

$$U_2(x_2,y_2) = \frac{\exp(-ikz)}{i\lambda z} \int\int_{-\infty}^{\infty} U_1(x_1,y_1) \cdot \exp\left[\frac{-ik}{2z}\left((x_2\text{-}x_2)^2 (y_2\text{-}y_1)^2\right)\right] dx_1dy_1 \qquad (2)$$

If $z \; >> \; \frac{1}{2} k \cdot (x_1{}^2+y_1{}^2)_{max}$ you can use the Fraunhofer approximation.

$$U_2(x_2,y_2) = \frac{\exp(-ikz)}{i\lambda z} \exp\left[\frac{-ik}{2z}(x_2{}^2+y_2{}^2)\right]$$

$$\cdot \int\!\!\int_{-\infty}^{\infty} U_1(x_1,y_1)\cdot \exp\left[\frac{ik}{z}(x_1x_2+y_1y_2)\right] dx_1dy_1 \ . \tag{3}$$

Equation (3) looks like a two-dimensional Fourier transform, which can be defined as:

$$\widetilde{U}(m,n) = \int\!\!\int_{-\infty}^{\infty} U(x,y)\cdot \exp\left[i2\pi\cdot(mx+ny)\right] dxdy \ . \tag{4}$$

The inverse relationship is

$$U(x,y) = \int\!\!\int_{-\infty}^{\infty} \widetilde{U}(m,n)\cdot \exp\left[-i2\pi(mx+ny)\right] dmdn \ . \tag{5}$$

So we can write equation (2) in the form

$$U_2(x_2,y_2)\cdot \exp\left[\frac{ik}{2z}(x_2{}^2+y_2{}^2)\right] = \frac{\exp(-ikz)}{i\lambda z}\,\widetilde{U}\left(\frac{x_2}{\lambda z}, \frac{y_2}{\lambda z}\right) . \tag{6}$$

In most practical applications the form of the limiting aperture is a circle. So for further discussion it is useful to calculate the Fraunhofer diffraction for a circular aperture. If a function is radially symmetric, then the two-dimensional Fourier transform is also radially symmetric and may be written as a Fourier Bessel transform.

$$\widetilde{U}(\rho) = \int_0^{\infty} U(r)\cdot J_0(2\pi\rho r)2\pi r dr \tag{7}$$

where J_n is a Bessel function of the first kind of order n. The diffracted amplitude for a circular aperture with the radius a is given by

$$U_2(r_2) \cdot \exp\left(\frac{ikr^2}{2z}\right) = \frac{\exp(-ikz)}{i\lambda z} \pi a^2 \cdot \frac{2J_1(2\pi r_2 a/\lambda z)}{2\pi r_2 a/\lambda z} \tag{8}$$

The two exponentional terms are only phase factors, which are cancelled by multiplying the amplitude with the complex conjugated term to get the intensity.

$$I_2(r_2) = \left(\frac{\pi a^2}{\lambda z}\right)^2 \cdot \left[\frac{2J_1(2\pi r_2 a/\lambda z)}{2\pi r_2 a/\lambda z}\right]^2 . \tag{9}$$

The factor ($\frac{\pi a^2}{\lambda z}$) is offten called the Fresnel number N.

In order to discuss the image formation it is necessary to calculate the influence of the imaging lens. For a basic discussion it is sufficient to consider a thin lens. By looking at equation (8), we can see that the phase change produced by the lens must be such as to multiply the amplitude by the factor

$$g(x,y) = \exp\left[\frac{ik}{2f}(x^2+y^2)\right] . \tag{10}$$

A real lens has a finite size. So, the pupil function P(x,y) has to be taken into account. If the lens is illuminated with a plane wave of unit strength the amplitude just after the lens can be written as

$$U(x,y) = P(x,y) \cdot \exp\left[\frac{ik}{2f}(x_2{}^2+y_2{}^2)\right] \cdot \tag{11}$$

Normally we are concerned with radially symmetric pupils. So the amplitude in a plane at a distance z away from the lens is given by

$$U_2(r_2) = \frac{\exp(-ikf)}{i\lambda f} \exp\left(\frac{-ikr_2^2}{2f}\right) \cdot \int_0^\infty P(r_1) J_0\left(\frac{2\pi r_1 r_2}{\lambda f}\right) 2\pi r_1 dr_1 . \tag{12}$$

The integral

$$h(r_2) = \int_0^\infty P(r_1) J_0\left(\frac{2\pi r_1 r_2}{\lambda f}\right) 2\pi r_1 dr_1 \tag{13}$$

is called the point spread function h(r) of the lens. By evaluating the integral for a radially symmetric pupil with the radius a, we get for the diffracted amplitude

$$U_2(v) = -iN \cdot \exp(-ikf) \cdot \exp\left(\frac{-iv^2}{4N}\right) \cdot \frac{2J_1(v)}{v} \tag{14}$$

Here the normalised optical coordinate v is introduced, which is defined by

$$v = kr_2 \sin\alpha \approx \frac{2\pi a}{\lambda f} r_2 \tag{15}$$

where $\sin\alpha$ is the numerical aperture of the lens.

We now have to discuss the effect of defocus. The amplitude in a plane at a distance δz away from the focal plane may be written as

$$U_2(v) = -iN \cdot \exp(-ikz) \cdot \exp\left(\frac{-iv^2}{4N}\right) \cdot \int_0^\infty 2\exp\left[\frac{ik\rho^2 a^2}{2}\left(\frac{1}{f} - \frac{1}{z}\right)\right] \cdot J_0(v\rho)\rho d\rho \tag{16}$$

with the definition of a second normalised optical coordinate u by

$$u = ka^2\left(\frac{1}{f} - \frac{1}{z}\right)$$

$$\approx k\frac{a^2}{f^2}\,\delta z \qquad \text{(for small } \delta z\text{)}$$

we get for the intensity along the optic axis

$$I_2(u,0) = N^2 \left(\frac{\sin(u/4)}{u/4}\right)^2 \tag{18}$$

In general the intensity is given by

$$I(u,v) = N^2 \cdot \left[C^2(u,v)+S^2(u,v)\right]^2, \tag{19}$$

where C(u,v) and S(u,v) defined as

$$C(u,v) = \int_0^1 2\cos\left(\frac{1}{2}u\rho^2\right) \cdot J_0(v\rho)\rho d\rho \tag{20}$$

$$S(u,v) = \int_0^1 2\sin\left(\frac{1}{2}u\rho^2\right) \cdot J_0(v\rho)\rho d\rho \tag{21}$$

With the formulas derivated above we are now able to calculate the image of a sample positioned in a plane (x_1,y_1) before a lens. Figure 3 shows the image formation geometry. The sample which lies in the (x_1,y_1) plane may be described by a complex amplitude transmittance $t(x,y)$. So if the sample is illuminated with a plane wave of unit strength the amplitude just behind the sample is similarly $t(x,y)$. To get the amplitude $U_2(x_2,y_2)$ immediately behind the lens we have to apply the Fresnel diffraction formula and multiply with the pupil function and phase factor of the lens. By a further application of equation (2) we get the amplitude $U_3(x_3,y_3)$ in a plane at a distance d_2 behind the lens.

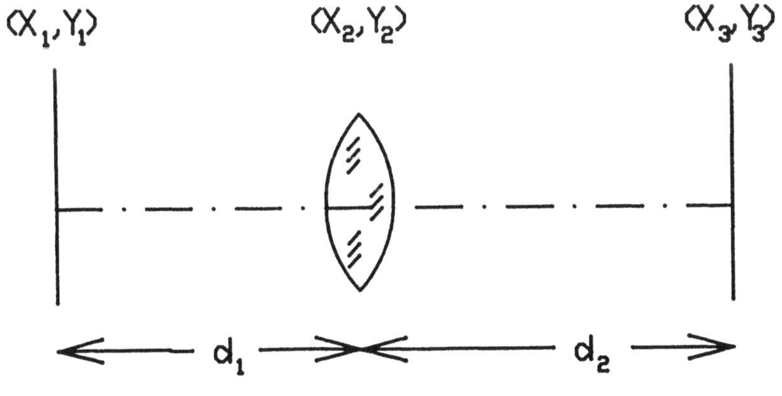

Fig.3: The image formation geometry.

$$U_3(x_3,y_3) = -\frac{\exp[-ik(d_2+d_2)]}{\lambda^2 d_1 d_2} \int\int_{-\infty}^{+\infty}\int\int P(x_2,y_2)\cdot t(x_1,y_1)\exp\left[\frac{-ik}{2d_1}(x_1{}^2+y_1{}^2)\right]$$

$$\cdot \exp\left[\frac{-ik}{2d_2}(x_3{}^2+y_3{}^2)\right]\cdot \exp\left[\frac{-ik}{2}\left(\frac{1}{d_1}+\frac{1}{d_2}-\frac{1}{f}\right)\cdot(x_2{}^2+y_2{}^2)\right]$$

$$\cdot \exp\left[ik\left(x_2\left(\frac{x_1}{d_1}+\frac{x_3}{d_2}\right)+y_2\left(\frac{y_1}{d_1}+\frac{y_3}{d_2}\right)\right]dx_1dy_1dx_2dy_2 \qquad (22)$$

When the lens law $\frac{1}{d_1}+\frac{1}{d_2}=\frac{1}{f}$ is satisfied, we perform the integral in x_2,y_2 to get the following expression:

$$U_3(x_3,y_3) = \frac{\exp(-ikd_1(1+M))}{\lambda^2 M d_1{}^2}\exp\left[\frac{-ik}{2Md_1{}^2}(x_3{}^2+y_3{}^2)\right] \qquad (23)$$

$$\cdot\int\int_{-\infty}^{\infty} t(x_1,y_1)\exp\left[\frac{-ik}{2d_1}\left(x_1{}^2+y_1{}^2\right)\right]\cdot h\left(x_1+\frac{x_3}{M},\ y_1+\frac{y_3}{M}\right)dx_1dy_1$$

where

$$M = \frac{d_2}{d_1} \quad \text{and}$$

$$h(x,y) = \int\int_{-\infty}^{\infty} P(x_2,y_2)\cdot\exp\left[\frac{ik}{d_1}\left(x_2x+y_2y\right)\right]dx_2dy_2 \quad , \qquad (24)$$

is the Fourier transform of the pupil function and called the point spread function of the lens.

For paraxial approximation the point spread function for an ideal lens is given by (solving equation 24):

$$h(r) = \frac{J_1\left(2\pi/\lambda\cdot r\cdot\sin\alpha\right)}{2\pi/\lambda\cdot r\cdot\sin\alpha} \qquad (25)$$

This is a very important equation for the following discussion. If the point spread function is known, it is simple to calculate the image formation of the microscope.

The main difference between a standard microscope and a CSOM is the way of illumination and detection. In a standard microscope the image is detected by a large-area detector. For illumination normally an incoherent light source is used. Thus, any phase variation in the image will be averaged out, and to calculate the image only the intensities of two neighbouring points have to be taken into account. The intensity point spread function of a standard microscope is simply given by $|h(r)|^2$. The intensity transmittance of the object is given by $|t(r)|^2$. So, the intensity at any point of the image is given (looking at equation (23)) by the convolution of the intensity point spread function with the intensity transmittance of the sample.

$$I(r) = |h(r)|^2 \otimes |t(r)|^2 , \tag{26}$$

where $\otimes$ denotes the convolution.

In the CSOM at any time only one point is illuminated. The amplitude at the object is therefore given by the point spread function of the objective lens h(r). The same lens images the sample onto a point detector. So the amplitude point spread function of a confocal microscope is give by $h(r)^2$. The image amplitude is the convolution of the amplitude point spread function with the amplitude transmittance t(r). By squaring this convolution we get the image intensity of the CSOM:

$$I(r) = |h(r)^2 \otimes t(r)|^2. \tag{27}$$

Thus, for simple objects the image intensity of the CSOM is just the square of the image intensity of a conventional microscope. If we take a δ-like transmittance the convolution can be easily done. Accordingly, we get for the image intensity of a standard microscope

$$I(v) = \left[\frac{2J_1(v)}{v} \right]^2 , \tag{28}$$

and for the CSOM

$$I(v) = \left[\frac{2J_1(v)}{v} \right]^4 \tag{29}$$

If we want to discuss the resolution of these two different types of microscopes we have to ask at which distance two points can be separated. It is often usual to use the Rayleigh criterion: Two points are separated if the maximum of one falls upon the first minimum of the other. The first minimum of the Bessel function $J_1(v)$ is given at value v=3.8. So the distance at which two points can be separated is

$$r = v \frac{\lambda}{2\pi} \sin\alpha \approx 0.6 \cdot \lambda \ ,$$

if we take the numerical aperture $\sin\alpha = 1$.

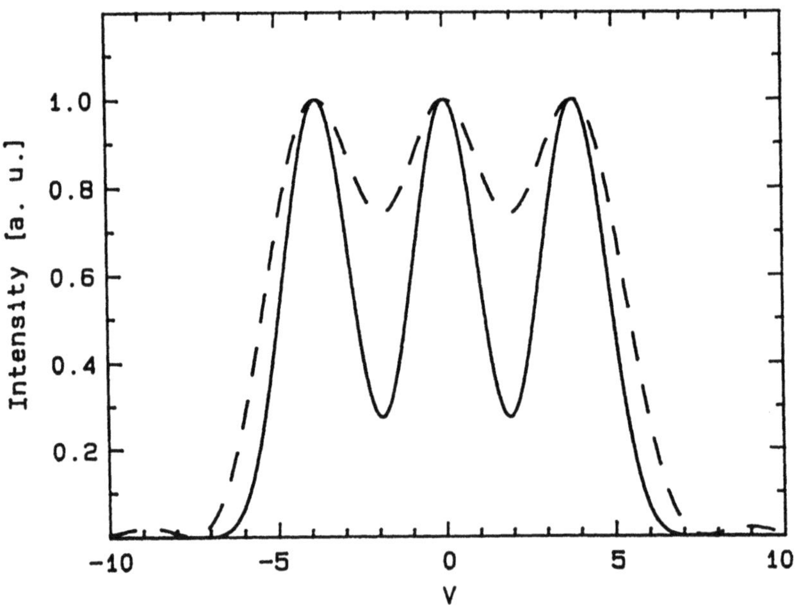

Fig.4: Image of 3 δ-like points with a spacing of v=3.8 for a normal microscope (dashed line) and a CSOM (full line).

If we use the Rayleigh criterion the CSOM will not lead to a better resolution. In figure 4 the image intensity for 3 δ-like points at a distance v=3.8 are plotted for a standard microscope and a CSOM. As we can see, the intensity modulation for the CSOM is substantially larger than that for the standard microscope. We can therefore define resolution as the drop of the image intensity to a fixed value. Now the distance v is reduced to this value as the intensity falls to 75% of the maximum. The result is plotted in figure 5. What we can see is that for the same intensity modulation as for a normal microcscope a CSOM can resolve two points whose distance is 30% closer.

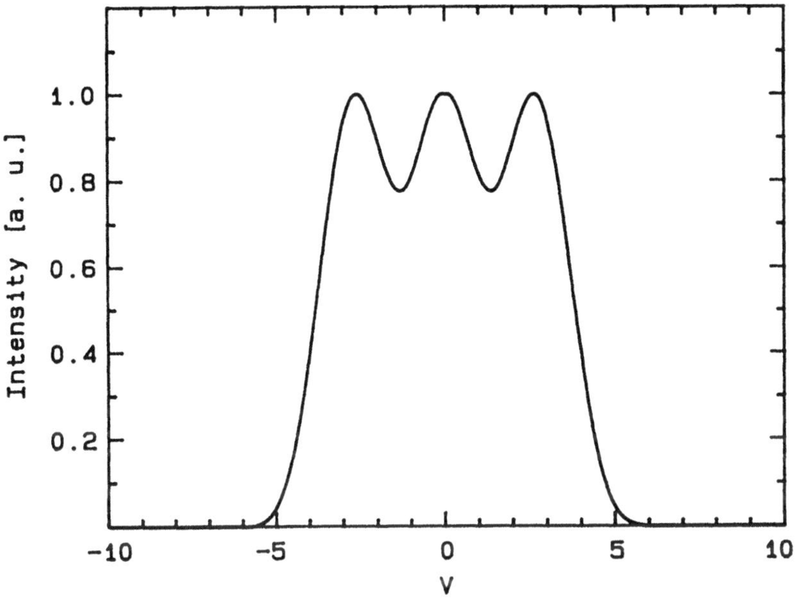

Fig.5: Image of 3 δ-like points with a spacing of v=2.7 for a CSOM.

A substantial difference between a conventional microscope and a confocal microscope is the real depth resolution of the CSOM. For a given object position in the (x,y)-plane the axial intensity is given by (see eq. 18),

$$I_2(u,0) = N^2 \cdot \left(\frac{\sin(u/4)}{u/4}\right)^2 \qquad (30)$$

for the standard microscope, and

$$I_2(u,0) = N^2 \cdot \left(\frac{\sin(u/4)}{u/4}\right)^2 \tag{31}$$

for the CSOM. This difference does not seem to be very large. But for practical importance the integrated intensity is more relevant.

$$I_{int}(u) = 2\pi \int_0^\infty I(u,v)v\,dv \tag{32}$$

This tells us how our microscope discriminates light that does not come out of the focal plane.

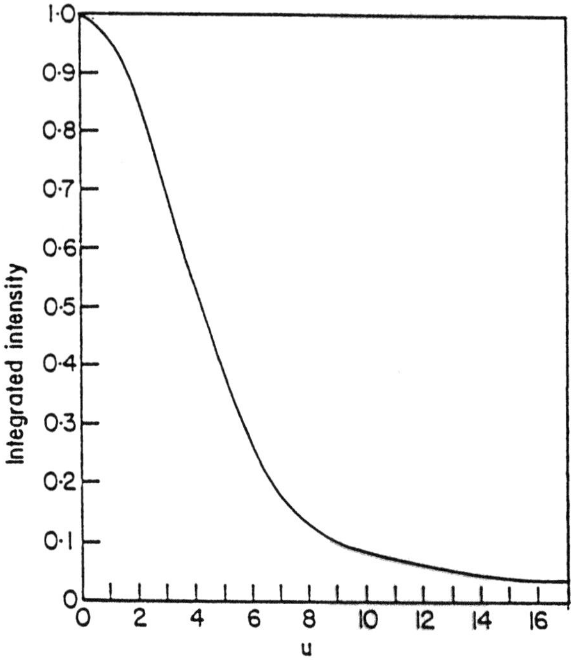

Fig.6: Integrated intensity for a CSOM.

For a conventional microscope the intensity $I(u,v)$ is the same as the intensity near the focus of the lens. So the integrated intensity is proportional to the power crossing any plane perpendicular to the optical axis, and according to the law of the conservation of energy this has to be constant. Therfore this type of microscope has no depth discrimination. In the case of a confocal microscope we have to solve equation (31) for the intensity given by equation (19). The result is plotted in figure 6. The integrated intensity falls to the half power point at a distance of $0.7 \cdot \lambda$ from the focal plane for an objective with a numerical aperture of unity. So the CSOM has a real depth resolution in the order of the wavelength.

IV. THE REAL-TIME CONFOCAL MICROSCOPE

There are different designs of confocal scanning microscopes described in the literature [7, 8]. If you want to get a complete image of the object you have to scan either the sample or the point source. With an electronic recording system the image of each point is stored and a computer is used to reconstruct the complete picture. Figure 7 schematically shows the path of ray for a laser scanning microscope and the basic components of the electronic equipment (after ref. [9]).

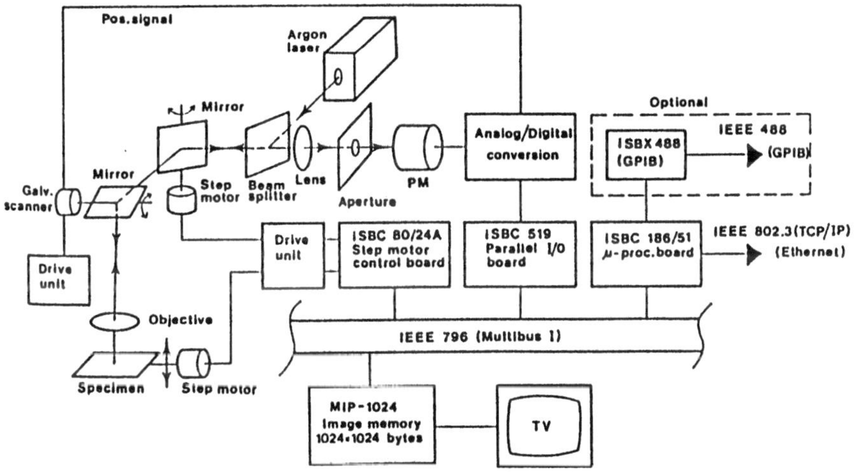

Fig.7: Schematic diagram of a confocal Laser scanning microscope used at the Royal Institute of Technology, Stockholm [9].

In the RSOM a rotating Nipkow disk instead of the single pinhole is used. The light source illuminates several hundred pinholes on the disk. Each point is imaged by the objective to a diffraction-limited spot on the sample. The light is reflected from the object and passes the pinholes on the Nipkow disk. The image can be directly seen with the eyepieces. It is not necessary to use any electronic recording. In this way it is possible to get a real-time image of the sample. Figure 8 shows the principle path of ray of our RSOM.

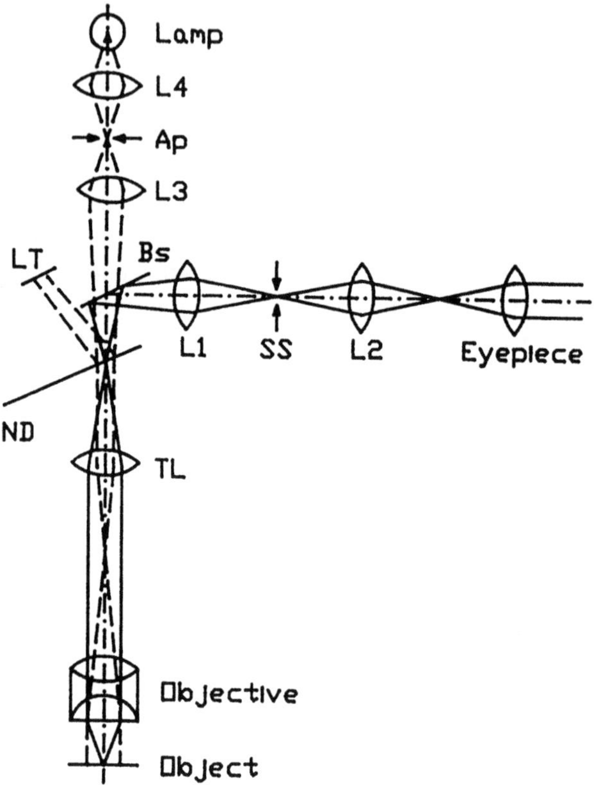

Fig.8: Schematic path of ray of our RSOM-design (L1...L4: lenses, Bs: beam-splitter, LT: light trap, ND: Nipkow disk, TL: tube lens, SS: stray light stop).

Since only 1% of the Nipkow disk is transparent it is important to reduce the stray light level to about 10^{-5} of the illumination intensity. In the literature [4, 10] descriptions can be found of various constructions using polarisation optics to reduce stray light. The disadvantage is that they include additional components in the optical arrangement which reduce the illumination intensity. Another possibility to reduce stray light is to introduce a stop in the path of ray, as shown in figure 8.

In our RSOM the first step to reduce stray light is to tilt the Nipkow disk by an angle ϑ_g against the perpendicular line to the optical axis (see fig. 8). As a result, the primary reflected beam cannot reach the detector. In addition to a normal incidental light microscope we have introduced a stray light stop in the pupillary intermediate plane. The diameter of the stop has to be exactly matched to the exit pupil image of the objective. A smaller diameter would reduce the resolution, caused by a decrease in the aperture of the microscope. A larger diameter would increase the stray light level. Because of the small amount of the angle ϑ_g we get with the sin-law

$$\sin(\vartheta_g) \approx \vartheta_g = \frac{NA}{M} \tag{33}$$

where M denotes the magnification and NA the numerical aperture of the objective. If we take NA=0.95 and M=160 for the used objective we get $\vartheta_g=5.9 \cdot 10^{-3}$.

From a spherical stray light distribution a small part, the dashed one shown in figure 9, travels along the optical axis in the cone denoted by a. So only the small part of the stray light scattered in the solid angle $\Omega_g=2\pi(1-\cos\vartheta_g)$ $=1.1 \cdot 10^{-4}$ sr can reach the detector. If we deal with a Lambert distribution instead of a spherical distribution for stray light, the reduction factor becomes $R=1.8 \cdot 10^{-5}$. Our experience shows that this reduction is sufficient to get a high quality image. A detailed calculation of the stray light reduction for the RSOM is given by B. Neumann [11].

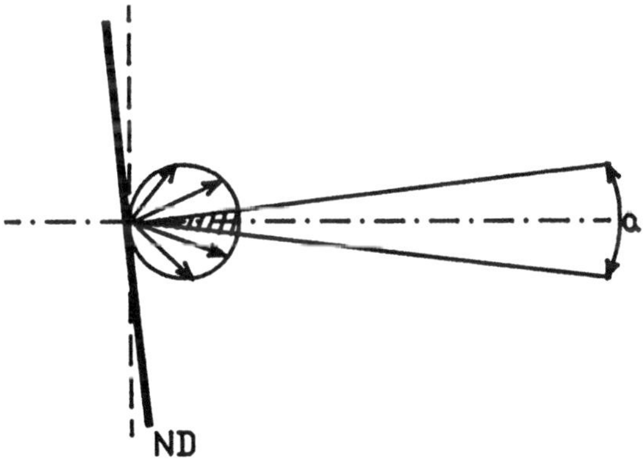

Fig.9: Only the dashed part of a spherical stray light distribution, generated by the Nipkow disk, can reach the detector.

The advantage of the RSOM in comparison with a standard microscope is demonstrated in figures 10 to 17 which show the photographs of the images of some samples. The first sample to be seen there shows a pyramid-like structure which was generated by pressing a diamond tip onto a metal surface.

By using a normal microscope we get the image of the structure as shown in figure 10. You can see a sharp surface and a diffuse background in the position where the structure might be. If we switch on the RSOM we are able to look into the structure. Figures 11 to 13 show different depth sections. We are going deeper into the pyramid-like hole up to that point where we reach the tip of the pyramid (fig. 13). By storing the different z-positions of the images of the sample, it is possible to measure the depth of the structure.

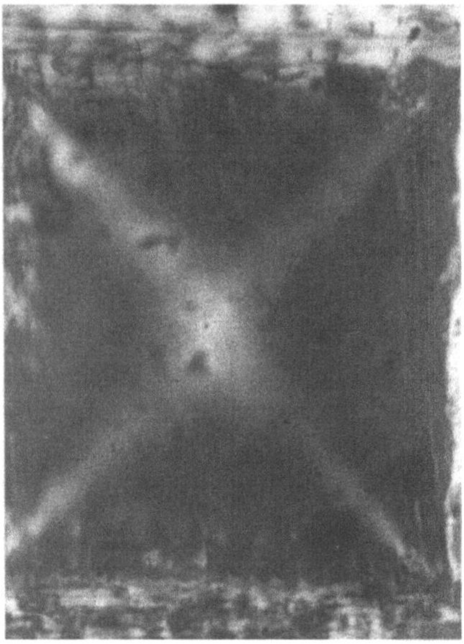

Fig.10: Photograph of a pyramid-like structure taken with a conventional microscope.

Fig.11: Same as in Fig.10, but taken with the RSOM. The focal plane lies at the bottom of the pyramid.

Figure 14 shows the image of an integrated circuit using a normal microscope. If we want to measure the line width of the horizontal lines, the reflected light from the vertical lines which lie in deeper sections will produce a constant background, which disturbes the image. By using the RSOM it is possible to look either at the horizontal lines or at the vertical lines, because these lie in different depth scections. The series of photographs (fig. 15 to 17) shows the structure of an integrated circuit, coming from the top and going deeper into the structure. In figure 17 the region of the metal contacts of the source and drain is clearly resolved. As you can see, the RSOM is an ideal measuring tool to examine samples with an interesting z-structure.

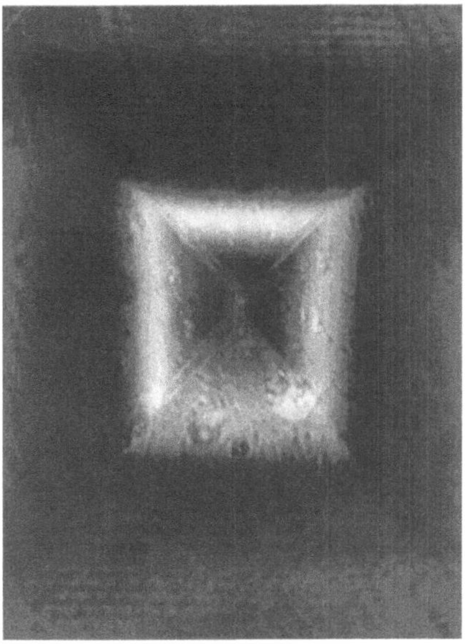

Fig.12: As fig. 11, but focal plane in the middle of the structure.

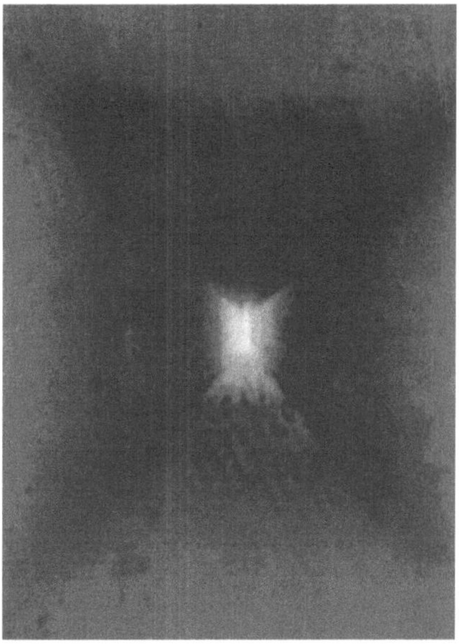

Fig.13: As fig. 12, but now the focal plane lies in the tip of the pyramid.

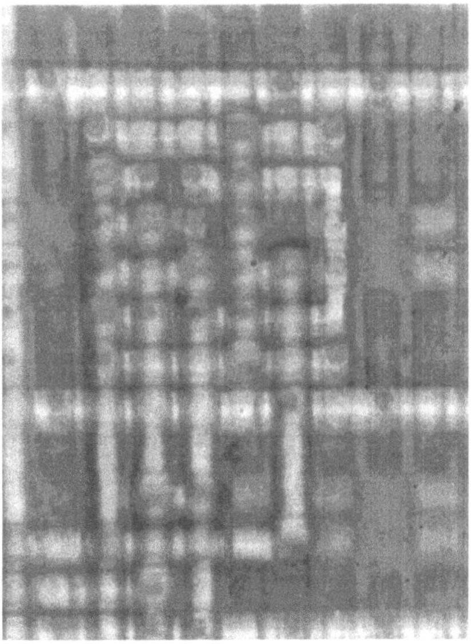

Fig.14: Photograph of an integrated circuit taken with a normal microscope.

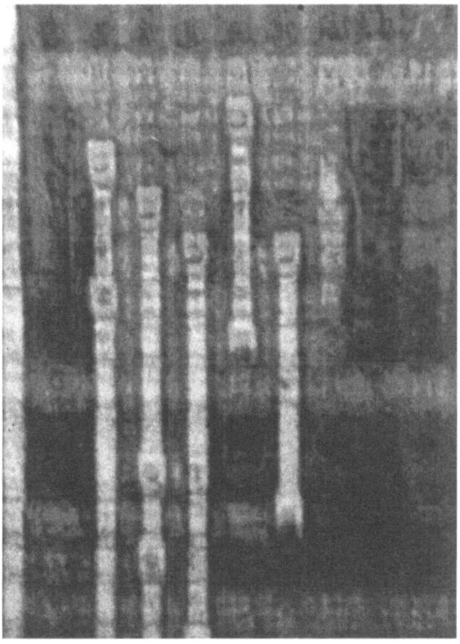

Fig.15: Same as fig. 14, but taken with the RSOM.

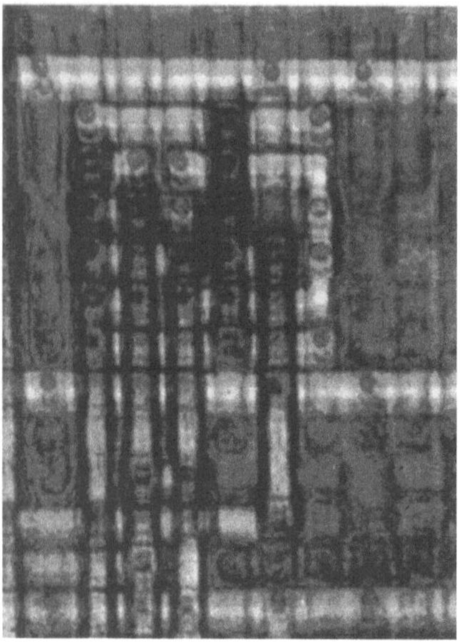

Fig.16: As fig. 15, but a different depth section.

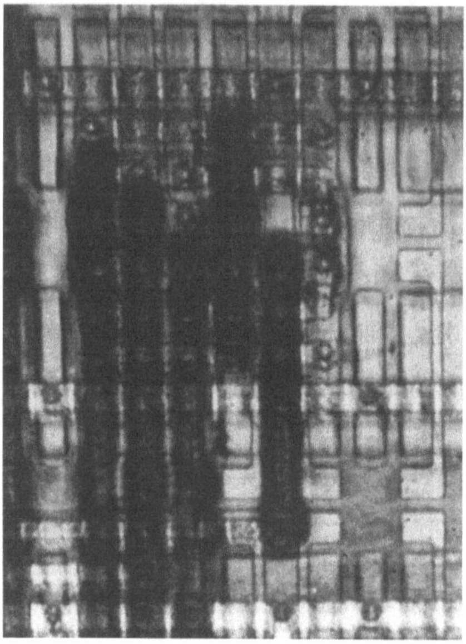

Fig.17: Another depth section of the integrated circuit.

V. COMPARISON WITH OTHER MICROSCOPES

The RSOM is one type of a large class of different scanning microscopes. The best known is the scanning electron microscope SEM. There are some other scanning microscopes, the scanning tunneling microscope STM and the scanning acustic microscope SAM, for example.

The great advantage of the SEM is the better definition in comparison with the CSOM. The typical resolution is in the order of 10 nm for electrons with 2 keV energy and 2.5 nm for 10 keV electrons. The disadvantage is the great expense to prepare the sample. In order to prevent the sample from getting negatively charged it first has to be coated with a conductor. Secondly, the sample has to stay in vacuum, and as it takes time to pump down the vacuum chamber, it is not possible to do insitu measurements. To obtain some depth information the sample has to be cross-sectioned, which destroys it for further use. That is another disadvantage.

The resolution of a STM is typically in the order of 1 or 2 Å and therefore ideal to study the atomic structure of the surface. Because of the limited dynamic range of the mechanical positioning systems it is difficult to examine a large area in a justifiable time.

The SAM has a resolution in the order of the acoustic wavelength in the material. For a frequency of 3 GHz the definition is in the order of 0.4 μm. The advantage of the acustic microscope is that it can examine optically opaque materials. The disadvantage is that the sample has to be immersed in water which is usually used as the coupling medium.

For special applications the different scanning microscopes feature some advantages in comparison with the RSOM. But optical systems have the great advantage of parallel processing which makes them suitable for getting a real-time image. This has great advantages for insitu measurements in semiconductor fabrication lines or for studying dynamic processes, for example in biological specimens. So, resolution is not everything, the RSOM has many advantages which make it a powerful measuring tool for the future.

VI. CONCLUSION

The great advantage of the confocal scanning microscope is the real depth resolution which is in the order of the wavelength. Additionally, the lateral resolution is 30% better than for a conventional microscope, caused by the increased contrast ratio. With the RSOM it is possible to view directly different depth sections of a sample without need of electronic recording. The stray light level can be effectively reduced by introducing a stop in the pupillary intermediate plane. The advantages of the RSOM make it possible to image a wide range of materials and to measure their properties.

References

1 J. Z. Young and F. Roberts, Nature **167**, 231 (1951)
2 M. Minsky, US patent 3013346, 19 December 1961
3 P. Davidovits and M. D. Egger, Nature **233**, 831 (1969)
4 G. Q. Xiao and G. S. Kino, A Real-time Confocal Scanning Optical Microscope, in SPIE Proc. **809**, Scanning Imaging Technology (1987)
5 M. Born and E. Wolf, Principle of Optics, (Pergamon New York 1983)
6 T. Wilson and C. Sheppard, Theory and Practice of Scanning Optical Microscopy (Academic, London 1984)
7 M. Petran, M. Hadravsky and A. Boyde, The Tandem Scannig Reflected Light Microscope, Scanning **7**, 125 (1985)
8 G. Q. Xiao, T. R. Corle and G. S. Kino, Appl. Phys. Lett. **53**, 716 (1988)
9 K. Carlson and N. Åslund, Applied Optics, **26**, 3232 (1987)
10 T. R. Corle, G. Q. Xiao, G. S. Kino and N. S. Levine, Characterization of a Real-time Confocal Scanning Optical Microscope, SPIE Proc. **1087**, Integrated Circuit Metrology, Inspection and Process Control III, 138 (1989)
11 B. Neumann, Optik, **85**, 11 (1990)

On the Search for Last Frontiers
Scanning Tunneling Microscopy and Related Techniques (Abstract)

Dr. D.W. Pohl
IBM Rüschlikon
Forschungslab. Zürich
Säumerstr. 4
CH-8803 Rüschlikon

Every child can tell when two pieces of solid material get in touch with each other - it is one of the most elementary experiences in human life. Popular criteria are that *there is no gap left, they can't be moved further towards each other* and, for conductors, *you can run a current from one to the other*. For the physicist/chemist at the end of the 20th century, the question is more difficult to answer. Having got used to atomic size dimensions, he perceives the border of a body not as a mathematical plane but rather as a transition zone, and therefore he is not even able to tell exactly "where" the surface is let alone when it touches that of a second body.

Still, physicists and chemists are fascinated about this zone since it is the seat of many important properties of a material. But even more significant, with the capability to "see" a surface (more generally, a whole body) atom by atom an ultimate limit for material characterization will be reached: Any chemical and physical property can be predicted, at least in principle, from this knowledge, including materials composing the vivid world. Any further progress in resolution, on the other hand, would mean to dive into the realm of subatomic structures (core electrons and nucleus) which are of little relevance for the physical and chemical properties of a material.

The interest and necessity to gain better insight into the structure of matter has been a major motivation for the development of higher and higher resolution microscopes. Conventional optical microscopes gave access to the regime of micrometer (10^{-6} *m*) dimensions. The younger electron microscopy opened the door to the nanometer (10^{-9} *m*) regime which is not far from the size of an atom. With scanning tunneling microscopy (STM), the ultimate limit appears to have been touched more closely than ever before.

The principle of STM is extremely simple: A macroscopic stylus is glided across an object surface at a distance of a few atomic diameters. The gap is kept at constant width by sensing a distance-dependent interaction, viz. the ability to drive a weak current across it. The current flow is a consequence by the *tunnel effect*, an important quantum mechanical phenomenon and extends over a very small distance range.

STM, in the past decade, has provided wealth of information on atomic and mesoscopic surface topographies beginning with the famous Si(111) surface and its reconstructions and ending with the roughness of diamond-turned disks and plates. Another promising directions aims at the imaging of organic and biological molecules such as DNA have been imaged in this way.

The principle of STM has been generalized recently by considering interactions ("X") different from tunneling. Examples are the interfacial forces, optical antenna effects, and even heat flow. These so-called SXM microscopies are sensitive to different materials' properties like chemical specificy.

STM and AFM extensions (Abstract)

Dr. H.K. Wickramasinghe
Thomas J. Watson Res. Center
IBM Res. Div.
P.O. Box 218
USA - Yorktown Heights NY 10598

As device sizes get smaller and smaller, conventional optical techniques for characterisation are reaching their limit. There is an ever increasing need to perform such measurements on the nanometer scale. The Scanning Tunneling Microscope and the Atomic Force Microscope has taught us how to position a tip with angstrom accuracy over a sample surface and scan it in X, Y and Z in order to form an image. It is possible to use this principle to image other physical properties of a surface than the density of electronic states provided by the Tunneling Microscope. In this talk, we shall review several extensions of the STM and AFM that we have demonstrated recently within our group. These include high resolution magnetic force imaging, c/v imaging, thermal imaging, photovoltage microscopy, chemical potential microscopy, optical and phonon absorption microscopy. The principles of operation of these microscopies will be described together with representative results obtained on various samples.
In the attractive mode force microscope, the forces are measured using a sensitive laser heterodyne probe. Force gradients down to 10^{-4} N/m and forces down to 10^{-12} N can be measured. In some of the other techniques, the idea of a tunneling thermometer is exploited, whereby it is possible to measure milli-degree thermoelectric variations down to the atomic scale. Photovoltage and c/v microscopy allows one to measure dopants and defects on a nanometer scale. Representative results from these different areas of nanometer will be presented.

Bibliography

Scanning Probe Microscopy

(Extract, papers published from 1990 on)

G. W. B. Schlüter

Helmut Hund GmbH
P.O. Box 21, D-6330 Wetzlar 21
Phone: +49 6441 200117 Facs: +49 6441 200122

The bibliography has been built up funded by the ESPRIT Basic Research Action 3200 OLDS and will be updated permanently. We therefore will be very grateful for comments, corrections or hints at publications not yet included.

The complete bibliography is available on request.

Abraham, D.L.; Veider, A., Schönenberger, Ch., Meier, H.P., Arent, D.J., Alvarado, S.F.; Nanometer resolution in luminescence microscopy of III-V-heterostructures; Appl.Phys.Lett. 56, 1564

Abraham, D.W.; McDonald, F.A.; Theory of magnetic force microscope images; Appl.Phys.Lett. 56, 1181

Akamine, S.; Barrett, R.C., Quate, C.F.; Improved atomic force microscope images using microcantilevers with sharp tips; Appl.Phys.Lett. 57, 316

Akamine, S.; Albrecht, T.R., Zdeblick, M.J., Quate, C.F.; A planar process for microfabrication of a scanning tunneling microscope; Sens.Act. A23, 964

Albrecht, T.R.; Akamine, S., Zdeblick, M.J., Quate, C.F.; Microfabrication of integrated scanning tunneling microscope; J.Vac.Sci.Technol.A 8, 317

Albrecht, T.R.; Akamine, S., Carver, T.E., Quate, C.F.; Microfabrictaion of cantilever styli for the atomic force microscope; J.Vac.Sci.Technol.A 8, 3386

Albrektsen, O.; Arent, D.J., Meier, H.P., Salemink, H.W.M.; Tunneling microscopy and spectroscopy of molecular beam epitaxy grown GaAs-AlGaAs interfaces; Appl.Phys.Lett. 57, 31

Anders, M.; Mück, M., Heiden, C.; Potentiometry for thin-film structures using atomic force microscopy; J.Vac.Sci.Technol.A 8, 394

Anders, M.; Mück, M., Heiden, C.; Potentiometrie mit dem Rasterkraftmikroskop; Verh.Dtsch.Phys.Ges. 4, 1108

Anselmetti, D.; Baumgartner, S., Wiesendanger, R., Güntherodt, H.-J.; Rastertunnelmikroskopie an Defektstrukturen im Kristall- und Ladungsdichtewellengitter von CDW-Systemen; Verh.Dtsch.Phys.Ges. 4, 1158

Anselmetti, D.; Geiser, V., Brodbeck, D., Overney, G., Wiesendanger, R., Güntherodt, H.-J.; Hexagonal and nonhexagonal superlattice structures on stage-1 alkali metal graphite intercalation compounds studied by scanning tunneling microscopy; Synth.Met. 38, 157

Anselmetti, D.; Geiser, V., Wiesendanger, R., Güntherodt, H.-J.; Oberflächenuntersuchungen an Stufe-1 Alkalimetall-Graphiteinlagerungsverbindungen mit dem Rastertunnelmikroskop; Verh.Dtsch.Phys.Ges. 4, 1156

Anselmetti, D.; Wiesendanger, R., Güntherodt, H.-J., Gruner, G.; Atomic-scale surface investigations of K0.3MoO3 (blue bronze) by scanning tunneling microscopy; Europhys.Lett. 12, 241

Arscott, P.G.; Bloomfield, V.A.; Scanning tunneling microscopy of nucleic acids and polynucletides; Ultramicroscopy 33, 127

Bai, C.; Dai, C., Zhu, C., Chen, Z., Huang, G., Wu, X., Zhu, D., Baldeswieler, J.D.; Scanning tunneling microscopy of silver containing salt of bis(ethylenedithio)tetrathiafulvalene; J.Vac.Sci.Technol.A 8, 484

Bakel, G.P.E.M. van; Hosson, J.T.M. De, Hibma, T.; Scanning tunneling microscopy imaging of transition-metal dichalcogenides; Appl.Phys.Lett. 56, 2402

Bando, H.; Kashiwaya, S., Tokumoto, S., Tokumoto, H., Anzai, H., Kinishita, N., Kajimura, K.; Tunneling spectroscopy on an orhanic superconductor (BEDT-TTF)2Cu(NCS)2; J.Vac.Sci.Technol.A 8, 479

Barrett, R.C.; Quate, C.F.; Imaging polished sapphire with atomic force microscopy; J.Vac.Sci.Technol.A 8, 400

Basaki, A.A.; Nogami, J., Quate, C.F.; Gallium growth and reconstruction on the Si(100) surface; J.Vac.Sci.Technol.A 8, 245

Baski, A.A.; Nogami, J., Quate, C.F.; Si(111)-5x1-Au reconstruction as studied by scanning tunneling microscopy; Phys.Rev.B 41, 10247

Becker, A.; Fries, Th., Wandelt, K., Kreibig, U., Schmid, G.; STM investigation of stabilized Au55-Clusters; J.Vac.Sci.Technol.A (preprint)

Becker, R.; Vickers, J.; Determination of surface atomic positions by scanning tunneling microscope observations; J.Vac.Sci.Technol.A 8, 226

Becker, R.S.; Higashi, G.S., Chabal, Y.J., Becker, A.J.; Atomic scale conversion of clean Si(111):H-1x1 to Si(111)-2x1 by electron-stimulated desorption; Phys.Rev.Lett. 65, 1917

Bedrossian, P.; Mortensen, K., Chen, D.M., Golovchenko, J.A.; Probing atomically inhomogeneous surfaces with tunneling microscopy; Nucl.Instrum.Meth.Phys.Res. B48, 296

Bendixen, C.; Besenbacher, F., Lægsgaard, E., Stensgaard, I., Thomsen, B., Westergaard, O.; Deoxyribonucleic acid structures visualized by scanning tunneling microscopy; J.Vac.Sci.Technol.A 8, 703

Berndt, R.; Gimzewski, J.K.; Scanning tunneling optical microscopy auf Silber-Nanostrukturen; Verh.Dtsch.Phys.Ges. 4, 1160

Bernhardt, R.H.; McGonigal, G.C., Schneider, R., Thomson, D.J.; Mechanisms for the deposition of nanometer-sized structures from organic fluids using the scanning tunneling microscope; J.Vac.Sci.Technol.A 8, 667

Bernhardt, R.H.; McGonigal, G.C., Thomson, D.J.; The effect of short duration voltage pulses on tunneling tips; J.Vac.Sci.Technol.A (1991) (preprint)

Berthe, R.; Hartmann, U., Heiden, C.; Scanning tunneling microscopy of the Abrikosov flux lattice with ferromagnetic probes; Appl.Phys.Lett. 57, 2351

Besenbacher, F.; Mortensen, K.; Recent studies by scanning tunneling microscopy; Europhys.News 21, 68

Biegelsen, D.K.; Bringans, R.D., Northrup, J.E., Swartz, L.-E.; Reconstructions of GaAs(-1-1-1) surfaces observed by scanning tunneling microscopy; Phys.Rev.Lett. 65, 452

Biegelsen, D.K.; Bringans, R.D., Northrup, J.E., Swartz, L.-E.; Surface reconstructions of GaAs(100) observed by scanning tunneling microscopy; Phys.Rev.B 41, 5701

Biegelsen, D.K.; Bringans, R.D., Northrup, J.E., Swartz, L.-E.; Reconstructions of GaAs(-1-1-1) surfaces observed by scanning tunneling microscopy; Phys.Rev.Lett. 65, 452

Biegelsen, D.K.; Swartz, L.-E., Bringans, R.D.; GaAs epitaxy and heteroepitaxy: a scanning tunneling microscopy study; J.Vac.Sci.Technol.A 8, 280

Biensan, P.; Roux, J.-C., Saadaoui, H., Flandrois, S.; STM of CrCl3-graphite intercalation compounds; Microsc.Microanal.Microstruct. 1, 103

Blackford, B.L.; Jericho, M.H.; Simple two-dimensional piezoelectric micropositioner for a scanning tunneling microscope; Rev.Sci.Instrum. 61, 182

Blackford, B.L.; Jericho, M.H.; A metallic replica/anchoring technique for STM or AFM imaging of large biological structures; (preprint)

Bocko, M.F.; The scanning tunneling microscope as a high-gain, low-noise displacement sensor; Rev.Sci.Instrum. 61, 3763

Bockris, J.O'M.; In situ STM studies of lead electrodeposition on graphite substrate; J.Electrochem.Soc. 137, 452

Boef, A.J. den; Preparation of magnetic tips for a scanning force microscope; Appl.Phys.Lett. 56, 2045

Boef, A.J. den; The influence of lateral forces in scanning force microscopy; Rev.Sci.Instrum. 62, 88 (1991)

Boland, J.J.; Villarrubia, J.S.; Identification of the products from the reaction of chlorine with the silicon(111)-(7x7) surface; Science 248, 838

Borbonus, M.; Haase, O., Koch, R., Rieder, K.H.; Untersuchungen des Rekonstruktionsverhaltens gestufter Au(110)-Oberflächen mit Rastertunnelmikroskopie und Elektronenbeugung; Verh.Dtsch.Phys.Ges. 4, 1157

Bräuer, S.; Pagnia, H., Sotnik, N., Urban, J., Wirth, W.; Wechselstrommessungen mit dem Rastertunnelmikroskop; BEDO 23, 323

Brodde, A.; Badt, D., Tosch, St., Neddermeyer, H.; The growth of Ag films on Si(100); J.Vac.Sci.Technol.A 8, 251

Brodde, A.; Wilhelmi, G., Wengelnik, H., Badt, D., Neddermeyer, H.; Rastertunnelmikroskopie an NI(100) und O/Ni(100); Verh.Dtsch.Phys.Ges. 4, 1156

Brown, A.; Cline, R.W.; A low cost, high performance imaging system for scanning tunneling microscopy; Rev.Sci.Instrum. 61, 1484

Brown, N.M.D.; Hong-Xing You; The observation of the structure of the natural oxide layer on a Cr(110) surface by scanning tunneling microscopy; Surf.Sci. 233, 317

Buchholz, S.; Oberflächenuntersuchungen mit Hilfe der Raster-Tunnelmikroskopie (Surface studies by scanning tunneling microscopy); GIT Fachz.Lab. 9, 1972

Bustamante, C.; Smith, S.B., Houseal, T.W., Maestre, M.F., Johnston, W., Tierney, B., Keller, D.J.; Manipulation and analysis of single DNA molecules on a microscope stage; Proc.Roy.Microsc.Soc. 25, S58

Carlsson, P.; Holmstrom, B., Kita, H., Uosaki, K.; STM charcteristics of n-GaAs and p-GaP in electrolyte solution; J.Electroanal.Chem. 283, 425

Carrejo, J.P.; Thundat, T., Nagahara, L.A., Lindsay, S.M., Majumdar, A.; Scanning tunneling microscopy investigations of polysilicon films under solution; (preprint)

Chen, C.J.; Origin of atomic resolution on metal surfaces in scanning tunneling microscopy; Phys.Rev.Lett. 65, 448

Chua, F.M.; Kuk, Y., Silverman, P.J.; Nucleation and growth of a monolayer oxide on Cu(110); J.Vac.Sci.Technol.A 8, 305

Ciraci, S.; Baratoff, A., Batra, I.P.; Tip-sample interaction effects in scanning-tunneling and atomic-force microscopy; Phys.Rev.B 41, 2763

Colchero, J.; Marti, O., Mlynek, J.; Ein kompaktes Lichtzeiger-Kraftmikroskop mit atomarer Auflösung; Verh.Dtsch.Phys.Ges. 4, 1106

Coratger, R.; Claverie, R., Ajustron, F., Beauvillain, J.; Scanning tunneling microscopy of defects by carbon bombardement on graphite surfaces; Surf.Sci. 227, 7

Cotterill, G.F.; Bartlett, R., Hughes, A.E., Sexton, B.A.; STM investigation of galena surfaces in air; Surf.Sci. 232, L211

Cox, G.; Szynka, D., Poppe, U., Graf, K., Urban, K., Kisielowski-Kemmerich, C., Krüger, J., Alexander, H.; ; Phys.Bl. 46, 350

Cutkosky, R.D.; Versatile scan generator and data collector for scanning tunneling microscopes; Rev.Sci.Instrum. 61, 960

Dagata, J.A.; Schneir, J., Harary, H.H., Evans, C.J., Postek, M.T., Bennett, J.; Modification of hydrogen-passivated silicon by scanning tunneling microscope operating in air; Appl.Phys.Lett. 56, 2001

Dagata, J.A.; Tseng, W., Bennett, J., Evans, C.J., Schneir, J., Harary, H.H.; Selective-area epitaxial growth of gallium arsenide on silicon substrates patterned using a scanning tunneling microscope operating in air; Appl.Phys.Lett. 57, 2437

Demuth, J.E.; Koehler, U., Hamers, R.J.; Surface diffractometry and lattice imaging of scanning tunneling microscopy images; J.Vac.Sci.Technol.A 8, 214

Denley, D.R.; Practical applications of scanning tunneling microscopy; Ultramicroscopy 33, 83

Denley, D.R.; Scanning tunneling microscopy of rough surfaces; J.Vac.Sci.Technol.A 8, 603

Dietz, P.; Herrmann, K.-H.; Scanning tunneling microscopy of organic films on conducting substrates; Surf.Sci. 232, 339

Dijkamp, D.; Loenen, E.J. van, Hoeven, A.J., Dieleman, J.; Scanning tunneling microscopy study of Si(001) and Si(110) surface structures resulting from different thermal cleaning treatments; J.Vac.Sci.Technol.A 8, 218

Dijkkamp, D.; Hoeven, A.J., Loenen, E.J. van, Lenssinck, J.M., Dielman, J.; Morphology and distrubution of atomic steps on Si(001) studied with scanning tunneling microscopy; Appl.Phys.Lett. 56, 39

Din Ping; Jackson, H.E., Reddick, R.C., Sharp, S.H., Warmack, R.J.; Photon scanning tunneling microscope study of optical waveguides; Appl.Phys.Lett. 56, 1515

Doyen, G.; Koetter, E., Vigneron, J.P., Scheffler, M.; Theory of scanning tunneling microscopy; Appl.Phys.A 51, 281

Dransfeld, K.; Atomare Bewegungen in der Tunnel- und Kraftmikroskopie; Phys.Bl. 46, 307

Driscoll, R.J.; Youngquist, M.G., Baldeschwieler, D.; Atomic-scale imaging of DNA using scanning tunneling microscopy; Nature 346, 294

Ducker, W.A.; Force measurement using an atomic force microscope; J.Appl.Phys. 67, 4045

Dürig, U.; Zuger, O., Pohl, D.W.; Observation of metallic adhesion using the scanning tunneling microscope; Phys.Rev.Lett. 65, 349

Edstrom, R.D.; Meinke, M.H., Xiuru Yang, Rui Yang, Evans, D.F.; Scanning tunneling microscopy of the enzymes of muscle glycogenolysis; Ultramicroscopy 33, 99

Ehrichs, E.E.; Lozanne, A.L. de; Etching of silicon (111) with the scanning tunneling microscope; J.Vac.Sci.Technol.A 8, 571

Eigler, D.M.; Schweizer, E.K.; Positioning single atoms with a scanning tunneling microscope; Nature 344, 524

Elings, V.B.; Edstrom, R.D., Meinke, M.H., Yang, X., Yang, R., Evans, D.F.; Direct observations of enzymes and their complexes by scanning tunneling microscopy; J.Vac.Sci.Technol.A 8, 652

Emch, R.; Clivaz, X., Taylor-Denes, C., Vaudaux, P., Descouts, P.; Scanning tunneling microscopy for studying the biomaterial-biological tissues interface; J.Vac.Sci.Technol.A 8, 655

Endo, T.; Yamada, Y., Sumomogi, T., Kohno, K., Kuwahara, K., Fujita, T., Morita, S.; Atomic resolution images of metal surfaces exposed in oil using a scanning tunneling microscope; J.Appl.Phys. 68, 2528

Endo, T.; Yamada, H., Sumomogi, T., Kuwahara, K., Fujita, T., Morita, S.; Characteristics of newly developed scanning tunneling microscopy/spectroscopy at low temperatures; J.Vac.Sci.Technol.A 8, 468

Endo, T.; Yamada, Y., Sumomogi, T., Kohno, K., Kuwahara, K., Fujita, T., Morita, S.; Atomic resolution images of metal surfaces exposed in oil using a scanning tunneling microscope; J.Appl.Phys. 68, 2528

Erber, T.; McGreer, K.A., Nowak, E.R., Wan, J.-C., Weinstock, H.; Onset of hysteresis measured by scanning tunneling microscopy; J.Appl.Phys. 68, 1370

Ezaki, T.; Enomoto, H., Baba, M., Ikeda, Y., Ogura, S., Kuramochi, E., Ozaki, H.; Angle-resolved tunneling between two atomic planes; J.Vac.Sci.Technol.A 8, 182

Fan, F.-R.F.; Bard, A.J.; Scanning tunneling microscopy and tunneling spectroscopy of the n-TiO2(001) surface; J.Phys.Chem. 94, 3761

Fischer, U.Ch.; Scanning near field optical microscopy; (preprint)

Fischer, U.Ch.; Resolution and contrast generation in scanning near field optical microscopy; Proc.NATO Summer School Erice 475 (Kluwer Acad.Publ.)

Frenken, J.W.M.; Hamers, R.J., Demuth, J.E.; Thermal roughening studied by scanning tunneling microscopy; J.Vac.Sci.Technol.A 8, 293

Fuchs, H; Akari, S., Dransfeld, K; Molecular resolution of Langmuir-Blodgett monolyaers on tungsten diselenide by scanning tunneling microscopy; Phys.B 80, 389

Fuchs, H.; Dransfeld, K.; Molecular resolution of Langmuir-Blodgett monolayers on tungsten diselenide by scanning tunneling microscopy; Z.Phys.B 80, 389

Fuchs, H.; Laschinski, R.; Surface investigations with a combined SEM-STM; Scanning 12, 126

Fuchs, H.; Laschinski, R., Schimmel, T.; Atomic resolution of nanometer scale plastic surface deformations by scanning tunneling microscopy; Europhys.Lett. 13, 307

Fujita, M.; Nagayoshi, H., Yoshimori, A.; Electronic-structure theory of the dimers adatoms and stacking fault model on Si(111) reconstracted surface - comparison with scanning tunneling spectroscopy; J.Vac.Sci.Technol.A 8, 166

Gallagher, M.C.; Adler, J.G.; Low temperature scanning tunneling topogtaphy of YBa2Cu3Ox superconducting films; J.Vac.Sci.Technol.A 8, 464

García Cantú, R.; Huerta Garnica, M.A.; Long-scan imaging by scanning tunneling microscopy; J.Vac.Sci.Technol.A 8, 354

Garnaes, J.; Kragh, F., Morch, K.A., Thölén, A.R.; Transmission electron microscopy of scanning tunneling tips; J.Vac.Sci.Technol.A 8, 441

Gilbert, S.E.; Kennedy, J.H.; STM tunneling spectroscopy of TiO2(001) in air; Surf.Sci. 225, L1

Gimzewski, J.K.; Moller, R., Myhra, S., Schlittler, R.R., Stoll, E., Riviere, J.C., Calka, A., Radlinski, A.P.; Investigations of the surface of the amorphous alloy Fe80B20 by STM, XPS and AES; J.Non-Cryst.Solids 116, 253

Girard, C.; Maghezzi, S., Labeke, D. van; Interaction between a spherical probe and an atomic lattice: implication for atomic force microscopy on graphite and diamond; Surf.Sci. 234, 181

Göddenhenrich T.; Lemke, H., Mück, M., Hartmann, U., Heiden, C.; Probe calibration in magnetic force microscopy; Appl.Phys.Lett. 57, 2612

Göddenhenrich, T.; Lemke, H., Hartmann, U., Heiden C.; Rasterkraftmikroskopie mit kapazitivem Ausleseverfahren; Verh.Dtsch.Phys.Ges. 4, 1107

Göddenhenrich, T.; Lemke, H., Hartmann, U., Heiden, C.; Force microscope with capacitive displacement detection; J.Vac.Sci.Technol.A 8, 383

Göddenhenrich, T.; Lemke, H., Hartmann, U., Heiden, C.; Magnetic force microscopy of domain wall stray fields on single-crystal iron whiskers; Appl.Phys.Lett. 56, 2578

Gómez-Herrero, J.; Gómez-Rodríguez, J.M., García, R., Baró, A.M.; High values in scanning tunneling microscopy: field emission and tunnel regimes; J.Vac.Sci.Technol.A 8, 445

Gomez-Rodriguez, J.M.; Baro, A.M., Parkhutik, V.P.; Morphology of porous silicon studied by STM/SEM; Appl.Surf.Sci. 44, 185

Gómez-Rodríguez, J.M.; Baró, A.M., Silveira, J.P., Vázquez, M., González, Y., Briones, F.; Observation of AlGaAs/GaAs multiquantum well structure by scanning tunneling microscopy; Appl.Phys.Lett. 56, 36

Gould, S.A.C.; Drake, B., Prater, C.B., Weisenhorn, A.L., Manne, S., Hansma, H.G., Hansma, P.K., Massie, J., Longmire, M., Elings, V., Dixon Northern, B., Mukergee, B., Peterson, C.M., Stoeckenius, W., Albrecht, T.R., Quate, C.F.; From atoms to integrated circuit chips, blood cells, and bacteria with atomic force microscope; J.Vac.Sci.Technol.A 8, 369

Gould, S.A.C.; Drake, B., Prater, C.B., Weisenhorn, A.L., Manne, S., Keldermann, G.L., Butt, H.-J., Hansma, H., Hansma, P.K., Magonov, S., Cantow, H.J.; The atomic force microscope: a tool for science and industry; Ultramicroscopy 33, 93

Grafström, S.; Kowalski, J., Neumann, R., Probst, O., Wörtge, M.; Untersuchungen und Kompensation thermischer Effekte bei Einstrahlung von Laserlicht auf Probe und Spitze eines Rastertunnelmikroskops; Verh.Dtsch.Phys.Ges. 4, 1159

Grafström, S.; Kowalski, J., Neumann, R., Probst, O., Wörtge, M.; A compact scanning tunneling microscopy control and data acquisition system based on a Macintosh II workstation; J.Vac.Sci.Technol.A 8, 357

Grafstrom, S.; Kowalski, J., Neumann, R.; Design and detailed analysis of a scanning tunneling microscope; Meas.Sci.Technol. 1, 139

Grütter, P.; Jung, Th., Heinzelmann, H., Wadas, A., Meyer, E., Hidber, H.-R., Güntherodt, H.-J.; 10-nm resolution by magnetic force microscopy on FeNdB; J.Appl.Phys. 67, 1437

Grütter, P.; Rugar, D., Mamin, H.J., Castillo, G., Lin, C.-J., Valletta, R.M., Wolter, O., Bayer, T., Greschner, J.; Batch fabricated sensors for magnetic force microscopy; Appl.Phys.Lett. 57, 1820

Grütter, P.; Wadas, A., Meyer, E., Heinzelmann, H., Hidber, H.-R., Güntherodt, H.-J.; High-resolution magnetic force microscopy; J.Vac.Sci.Technol.A 8, 406

Grütter, P.; Wadas, A., Meyer, E., Heinzelmann, H., Hidber, H.-R., Güntherodt, H.-J.; High-resolution magnetic force microscopy; J.Appl.Phys. 67, 5953

Guinea, F.; Garcia, N.; Scanning tunneling microscopy, rsonant tunneling, and counting electrons: a quantum standard of current; Phys. Rev. Lett. 65, 281

Haase, J.; Hillert, B., Bradshaw, A.M., Kuk, Y., Silverman, P.J., Chua, F.M.; Comment on 'Oxygen chemisorption on Cu(110): an atomic view by scanning tunneling microscopy'; Phys.Rev.Lett. 64, 3098

Haase, O.; Borbonus, M., Muralt, P., Koch, R., Rieder, H.; A novel ultrahigh vacuum scanning tunneling microscope for surface science studies; Rev.Sci.Instrum. 61, 1480

Haberern, K.W.; Pashley, M.D.; GaAs(111) A-(2x2) reconstruction studied by scanning tunneling microscopy; Phys.Rev.B 41, 3226

Hallmark, V,M.; Chiang, S., Gölzhäuser, A., Killinger, A., Wöll, Ch.; Die Adsorption von Naphtalin auf Pt(111): Bestimmung von Orientierung und Anordnung der Moleküle mit dem Rastertunnelmikroskop; Verh.Dtsch.Phys.Ges. 4, 1157

Hameroff, S.; Simic-Krstic, Y., Vernetti, L., Lee, Y.C., Sarid, D., Wiedmann, J., Elings, V., Kjoller, K., McCuskey, R.; Scanning tunneling microscopy of Cytoskeletal proteins: microtubules and intermediate filaments; J.Vac.Sci.Technol.A 8, 687

Hamers, R.J.; Cahill, D.G.; Ultrafast time resolution in scanned probe microscopies; Appl.Phys.Lett. 57, 2031

Hamers, R.J.; Köhler, U.K., Demuth, J.E.; Epitaxial growth of silicon on Si(001) by scanning tunneling microscopy; J.Vac.Sci.Technol.A 8, 195

Hamers, R.J.; Market, K.; Surface photovoltage on Si(111)x(7x7) probed by optically pumped scanning tunneling microscope; J.Vac.Sci.Technol.A 8, 3524

Harada, S.; Goda, M.; Scanning tunneling microscopy images of carburized nickel alloys; J.Vac.Sci.Technol.A 8, 308

Hara, M.; Iwakabe, Y., Tochigi, K., Sasabe, H., Garito, A.F., Yamada, A.; Anchoring structure of smectic liquid-crystal layers on MoS2 observed by scanning tunneling microscopy; Nature 344, 228

Hartmann, U.; Theory of magnetic force microscopy; J.Vac.Sci.Technol.A 8, 411

Hartmann, U.; Van der Waals-Kräfte in der Rasterkraftmikroskopie; Verh.Dtsch.Phys.Ges. 4, 1107

Hartmann, U.; Magnetic microfield analysis by force microscopy; J.Magn.Magn.Mat. 83, 545

Hasegawa, T.; Kitazawa, K.; Atomically resolved STM/STS observation on the edge of CuO2 and (BiO)2 layers of Bi-Sr-Ca-Cu-O; Jpn.J.Appl.Phys. 29, L434

Hasegawa, T.; Takata, K., Hosaka, S., Hosoki, S.; Air-induced reconstruction of the Si(111) surface; J.Vac.Sci.Technol.A 8, 241

Hasegawa, Y.; Kamiya, I., Hashizume, T., Sakurai, T., Tochihara, T., Kubota, M., Murata, Y.; Adsorption of Li(K) on the Si(001)-(2x1) surface: scanning-tunneling-microscopy study; Phys.Rev.B 41, 9688

Hasegawa, Y.; Kamiya, I., Hashizume, T., Sakurai, T., Tochihara, H., Kubota, M., Murata, Y.; Cluster formation of Li in the Si(111) 7x7 surface; J.Vac.Sci.Technol.A 8, 238

Hashizume, T.; Hamers, R.J., Demuth, J.E., Markert, K., Sakurai, T.; Initial stage deposition of Ag on the Si(100)2x1 surface studied by scanning tunneling microscopy; J.Vac.Sci.Technol.A 8, 249

Hashizume, T.; Hasegawa, Y., Kamiya, I., Ide, T., Sumita, I., Hyodo, S., Sakurai, T., Tochihara, H., Kubota, M., Murata, Y.; Field ion-scanning tunneling microscopy of alkali metal adsorption on the Si(100) surface; J.Vac.Sci.Technol.A 8, 233

Hashizume, T.; Sajurai, T.; Alkali metal adsorption on the Si surfaces studied by the FI-STM (in Japanese); Sol.State Phys.(Jpn.) 25, 467

He Jiang; Wang Lanping, Chang Yue, Wang Guowen; Magnetically controlled coarse approach device for STM; Meas.Sci.Technol. 1, 1116

Heckl, W.M.; Ohnesorge, F., Egger, M., Weisenhorn, A.L., Specht, M., Schindler, H., Gaub, H.E., Hansma, P.K.; Rastertunnel- und kraftmikroskopische Messungen an organischen Proben am Beispiel von Lipid- und Proteinmembranen; Verh.Dtsch.Phys.Ges. 4, 1159

Heckl, W.M.; Specht, M., Möller, R., Ohnesorge, F.; Rastertunnel- und Kraftmikroskopie im Vergleich am Beispiel von Molybdänsulfid; Verh.Dtsch.Phys.Ges. 4, 1159

Hess, H.F.; Robinson, R.B., Dynes, R.C., Valles Jr., J.M., Waszczak, J.V.; Spectroscopy and spatial characterization of superconducting vortex core states with a scanning tunneling microscope; J.Vac.Sci.Technol.A 8, 450

He, J.; Lei, H.N., Kuang, Y.L., Yao, J.E.; SSX-1 STM high-speed microcomputer system for digital image display and processing; J.Vac.Sci.Technol.A (preprint)

Hoeven, A.J.; Dijkamp, D., Loenen, E.J. van, Lenssinck, J.M., Dieleman, J.; Island and step structures on molecular beam epitaxy grown Si(001) surfaces; J.Vac.Sci.Technol.A 8, 207

Hoeven, A.J.; Loenen, E.J. van, Hooft, P.J.G.M. van, Oostveen, K.; A multiprocessor data acquisition and analysis system for scanning tunneling microscopy; Rev.Sci.Instrum. 61, 1668

Hoffmann-Millack, B.; Roberts, C.J., Steer, W.S.; Surface modification and atomic resolution on a vacuum-annealed gold foil in air by scanning tunneling microscopy; J.Appl.Phys. 67, 1749

Hosaka, S.; Hasegawa, T., Hosoki, S., Takata, K.; Dynamic observation of Si(111) surface using a fast scanning tunneling microscope; Appl.Phys.Lett. 57, 138

Hosaka, S.; Sagara, K., Hasegawa, T., Takata, K., Hosoki, S.; Tunneling barrier hight imaging and polycrystalline Si surface observations; J.Vac.Sci.Technol.A 8, 270

Huang, Z.H.; Cutler, P.H., Feuchtwang, T.E., Kazes, E., Nguyen, H.Q., Sullivan, T.E.; Model studies of tunneling time; J.Vac.Sci.Technol.A 8, 186

Huang, Z.H.; Feuchtwang, T.E., Cutler, P.H., Kazes, E.; The Wentzel-Kramers-Brillouin method in multidimensional tunneling: application to scanning tunneling microscopy; J.Vac.Sci.Technol.A 8, 177

Humbert, A.; Scanning tunneling microscopy and spectroscopy; Arab.J.Sci.Eng. 15, 251

Humbert, A.; Dayez, M., Sangay, S., Chapon, C., Henry, C.R.; Scanning tunneling microscopy and transmission electron microscopy studies of Au and Pd clusters grown on a clean graphite surface; J.Vac.Sci.Technol.A 8, 311

Ibe, J.P.; Bey Jr., P.P., Brandow, S.L., Brizzolara, R.A., Burnham, N.A., DiLella, D.P., Lee, K.P., Marrian, C.R.K., Colton, R.J.; On the electrochemical etching of tips for scanning tunneling microscopy; J.Vac.Sci.Technol.A 8, 3570

Ichinokawa, T.; Ichinose, T., Tohyama, M., Itoh, H.; Scanning tunneling microscopy observation of MoS2 surface and gold clusters deposited on MoS2 surface; J.Vac.Sci.Technol.A 8, 500

In-Whan Lyo; Avouris, Ph.; Atomic scale desorption processes induced by scanning tunneling microscope; J.Chem.Phys. 93, 4479

Ishizaka, T.; Morita, S., Sugawara, Y., Okada, T., Mishima, S., Imai, S., Mikoshiba, N.; Surface imaging in air with a force microscope; J.Vac.Sci.Technol.A 8, 391

Ishizaka, T.; Sugawara, Y., Kumagai, K., Morita, S.; Anomalous force dependence of AFM corrugation height of a graphite surface in air; Jpn.J.Appl.Phys. 29, L1196

Itaya, K.; Sugawara, S., Sashikata, K., Furuya, N.; In situ scanning tunneling microscopy of platinum (111) surface with the observation of monatomic steps; J.Vac.Sci.Technol.A 8, 515

Jahanmir, J.; West, P.E., Colter, P.C.; Local modification of thin SiO2 films in a scanning tunneling microscope; J.Appl.Phys. 67, 7144

Jericho, M.H.; Blackford, B.L., Dahn, D.C., Frame, C., Maclean, D.; Scanning tunneling microscopy imaging of uncoated biological material; J.Vac.Sci.Technol.A 8, 661

Jobin, M.; Emcj, R., Zenhausern, F., Steinemann, S., Descouts, P.; Characterization of oxide film on titanium by scanning tunneling microscopy/spectroscopy: influence of the tip composition; J.Vac.Sci.Technol.A (1991) (preprint)

Johnson, M.; Clarke, J.; Spin-polarized scanning tunneling microscope: concept, design, and preliminary results from a prototype operated in air; J.Appl.Phys. 67, 6141

Kanchenko, V.A.; Krynko, Yu.N., Lyubinetsky, I.V., Melnik, P.V., Nakhodkin, N.G.; Use of scanning tunneling microscopy for the consideration of the sample surface roughness in spectroscopy of elastically reflected electrons (in Russian); Ukr.Fiz.Zh. 35, 1073

Kaneko, R.; Hamada, E.; Local modification of organic dye materials by dielectric breakdown; J.Vac.Sci.Technol.A 8, 577

Kato, T.; Tanaka, I.; A scanning tunneling microscopy/spectroscopy system for cross-sectional observations of epitaxial layers of semiconductors; Rev.Sci.Instrum. 61, 1664

Kato, T.; Tanaka, I., Sugiyama, N., Osaka, F.; Current imaging tunneling spectroscopy of thin n-GaAs/p-GaAs mulitlayer structures in air; Jpn.J.Appl.Phys. 29, L1189

Kawakatsu, H.; Higuchi, T.; A dual tunneling-unit scanning tunneling microscope; J.Vac.Sci.Technol.A 8, 319

Keller, R.W.; Dunlap. D.D., Bustamante, C., Keller, D.J., Garcia, R.G., Gray, C., Maestre, M.F.; Scanning tunneling microscopy images of metal-coated bacteriophages and uncoated, double-stranded DNA; J.Vac.Sci.Technol.A 8, 706

Kent, A.D.; Maggio-Aprile, I., Niedermann, Ph., Renner, Ch., Fischer, O.; Scanning tunneling potentiometry studies of Y1Ba2Cu3O7-x and gold thin films; J.Vac.Sci.Technol.A 8, 459

Khurana, A.; STM unravels the vortex core in type II superconductors; Physics Today 43, 17 (Jun. 1990)

Kim, H.S.; Zheng, Y.C., Bryant, P.J.; Real-space iamges of noble metal (Cu, Ag, Au) thin films by scanning tunneling microscopy; J.Vac.Sci.Technol.A 8, 314

Klitsner, T.; Becher, R.S., Vickers, J.S.; Observation of the effect of tip electronic states on tunnel spectra acquired with the scanning tunneling microscope; Phys.Rev.B 41, 3837

Kobayashi, K.; Isshiki, N., Tsukuda, M.; Simulation of scanning tunneling microscopy/spectroscopy based on electronic states theory (in Japanese); Sol.State Phys.(Jpn.) 25, 39

Kobayashi, K.; Tsukada, M.; Simulation of scanning tunneling microscope image based on electronic states of surface/tip system; J.Vac.Sci.Technol.A 8, 170

Kobayashi, K.; Tsukada, M.; Effect of tip atomic structure on scanning tunneling spectroscopy; Sol.State Commun. 74, 1187

Köhler, U.; Heuer, D., Jusko, O., Pietsch, G.J.; Epitaktisches Wachstum von Germanium auf Si(111)-Oberflächen: Eine Untersuchung mit dem Rastertunnel-Mikroskop; Verh.Dtsch.Phys.Ges. 4, 1158

Kojima, I.; STM & AES study of the relation between ion sputter induced roughness and depth resolution; J.Electr.Spectrosc. 50, c9,

Komarica, S.; Kopp, W.-U.; Einsatzmöglichkeiten des Raster-Tunnelmikroskops zur Oberflächenanalyse metallographischer Proben; BEDO 23, 315

Komiyama, M.; Kobayashi, J., Morita, S.; Structure of platinum ultrafine particles in Pt/C catalyst observed by scanning tunneling microscopy; J.Vac.Sci.Technol.A 8, 608

Kopelman, R.; Lewis, A., Lieberman, K.; Nanometer light source and molecular exciton microscopy; J.Lumin. 45, 298

Kordic, S.; Loenen, E.J. van, Dijkamp, D., Hoeven, A.J., Moraal, H.K.; Scanning tunneling spectroscopy on cleaved silicon pn junctions; J.Vac.Sci.Technol.A 8, 549

Kotomin, E.; Shluger, A., Causa, M., Dovesi, R., Ricca, F.; The adsorption of SiO molecules on MgO surfaces as a model for the silicon lever atomic force microscope (AFM); Surf.Sci. 232, 399

Krieger, W.; Suzuki, T., Völcker, M., Walther, H.; Generation of microwave radiation in the tunneling junction of a scanning tunneling microscope; Phys.Rev.B 41, 10229

Kuk, Y.; Becker, R.S., Silverman, P.J., Kochanski, G.P.; Optical interactions in the junction of a scanning tunneling microscope; Phys.Rev.Lett. 65, 456

Kuk, Y.; Chua, F.M., Silverman, P.J., Meyer, J.A.; O chemisorption on Cu(110) by scanning tunneling microscopy; Phys.Rev.B 41, 12393

Kuk, Y.; Silverman, P.J.; Scanning tunneling spectroscopy of metal surfaces; J.Vac.Sci.Technol.A 8, 289

Kuwabara, M.; Clarke, D.R., Smith, D.A.; Anomalous superperiodicity in scanning tunneling microscope images of graphite; Appl.Phys.Lett. 56, 2396

Labeke, D. van; Maghezzi, S., Girard, C., Vigoreux, J.M.; A theoretical model for the atomic force microscope; EMAG MICRO 89, Proc. Inst.Phys.Electr.Microsc.Anal. 1, 13

Lawunmi, D.; Payne, M.C.; Theoretical investigations of the scanning tunneling microscope image of graphite; J.Phys.,Cond.Matt. 2, 3811

Lemke, H.; Göddenhenrich, T., Hartmann, U., Heiden, C.; Herstellung von Nanosonden für die Rasterkraftmikroskopie; Verh.Dtsch.Phys.Ges. 4, 1107

Lieberman, K.; Harush, S., Lewis, A., Kopelman, R.; A light source smaller than the optical wavelength; Science 247, 59

Lindsay, S.M.; Sankey, O.F., Li, Y., Herbst, C., Rupprecht, R.; Pressure and resonance effects in scanning tunneling microscopy; J.Phys.Chem. 94, 4655

Loenen, E.J. van; Dijkkamp, D., Hoeven, A.J., Lenssinck, J.M., Dielman, J.; Evidence for tip imaging in scanning tunneling microscopy; Appl.Phys.Lett. 56, 1755

Loenen, E.J. van; Dijkamp, D., Hoeven, A.J., Lenssinck, J.M., Dieleman, J.; Nanometer scale structuring of silicon by direct identation; J.Vac.Sci.Technol.A 8, 574

Loo, B.H.; Liu, Z.F., Fujishima, A.; Scanning tunneling microscopic images of an azobenzene derivative differently deposited on highly oriented pyrolytic graphite surfaces; Surf.Sci. 227, 1

Lucas, A.A.; Theories for tunneling microscopy; Europhys.News 21, 63

Luo, C.; Zhu, C., Ruan, L., Huang, G., Dai, C., Cheng, Z., Bai, C., Su, Y., Xu, S., Lin, K., Baldeschwieler, J.D.; Scanning tunneling microscopy of the phosphatidylcholine bilayers; J.Vac.Sci.Technol.A 8, 684

Mamin, H.J.; Rugar, D., Lambert, S.E., Franco, L.P., Sanders, I.L., Yogi, T., Beaulieu, T.; Magnetic force microscopy of recording media; J.Appl.Phys. 67, 5953

Manne, S.; Butt, H.J., Gould, S.A.C., Hansma, P.K.; Imaging metal atoms in air and water using the atomic force microscope; Appl.Phys.Lett. 56, 1758

Marrian, C.R.K.; Colton, R.J.; Low-voltage electron beam lithography with a scanning tunneling microscope; Appl.Phys.Lett. 56, 755

Masai, J.; Sorin, T., Kondo, S.; Scanning tunneling microscopic immunoassay: a preliminary experiment; J.Vac.Sci.Technol.A 8, 713

McBridge, S.E.; Wetsel Jr., G.C.; Quantitative characterization of physical processes during nanometer surface modification; Appl.Phys.Lett. 57, 2782

McCord, M.A.; Awschalom, D.D.; Direct deposition of magnetic dots using a scanning tunneling microscope; Appl.Phys.Lett. 57, 2153

McGonigal, G.C.; Bernhardt, R.H., Thomson, D.J.; Imaging alkane layers at the liquid/graphite interface with the scanning tunneling microscope; Appl.Phys.Lett. 57, 28

McMaster, T.J.; Carr, H., Miles, M.J., Cairns, P., Morris, V.J.; Polypeptide structures imaged by scanning tunneling microscope; J.Vac.Sci.Technol.A 8, 648

McMaster, T.J.; Carr, H., Miles, M.J., Cairns, P., Morris, V.J.; Adsorption of liquid crystals imaged using scanning tunneling microscopy; J.Vac.Sci.Technol.A 8, 672

McMaster, T.J.; Miles, M.J., Morris, V.J., Wills, H.H.; Imaging biomolecules with the scanning tunneling microscope - tips for imaging at the sharp end of microscopy; Proc.Roy.Microsc.Soc. 25, S59

Meepagala, S.C.; Real, F., Reyes, C.B., Novoselskaya, A., Rong, Z., Wolf, E.L.; Compact scanning tunneling microscope with easy-to construct X-Z inertial sample translation; J.Vac.Sci.Technol.A 8, 3555

Menslage, J.; Robrock, K.H.; Magnetische Kraftmikroskopie: Mechanismen der Domänenbeobachtung anhand einer magnetischen Platte; Verh.Dtsch.Phys.Ges. 4, 1107

Mera, Y.; Yanagisawa, H., Uota, M., Maeda, K.; Scanning tip microscope for study of electrical inhomogeneity on submicron scale; J.Vac.Sci.Technol.A 8, 561

Meyer, E.; Heinzelmann, H., Brodbeck, D., Overney, G., Overney, R., Howald, L., Hug, H., Jung, T., Hidber, H.-R., Güntherodt, H.-J.; Atomic resolution on the surface of LiF(100) by atomic force microscopy; J.Vac.Sci.Technol.A (1991) (preprint)

Meyer, E.; Heinzelmann, H., Rudin, H., Güntherodt, H.-J.; Atomic resolution on LiF(001) by atomic force microscopy; Z.Phys.B 79, 3

Meyer, E.; Wiesendanger, R., Anselmetti, D., Hidber, H.R., Güntherodt, H.-J., Lévy, F., Berger, H.; Different response of atomic force microscopy and scanning tunneling microscopy to charge density; J.Vac.Sci.Technol.A 8, 495

Meyer, G.; Amer, N.M.; Optical-beam-deflection atomic force microscopy: the NaCl(001) surface; Appl.Phys.Lett. 56, 2100

Meyer, G.; Amer, N.M.; Simultaneous measurement of lateral and normal forces with an optical-beam-deflection atomic force microscope; Appl.Phys.Lett. 57, 2089

Meyer, J.; Visualisation of images from scanning probe microscopy; Europ.Microsc.Anal. 6, 27

Miles, M.; Biomolecular scrutiny by the STM; Physics World 10, 28

Miles, M.J.; McMaster, T., Carr, H.J., Tatham, A.S., Shewry, P.R., Field, J.M., Belton, P.S., Jeenes, D., Hanley, B., Whittam, M., Cairns, P., Morris, P., Lambert, N.; Scanning tunneling microscopy of biomolecules; J.Vac.Sci.Technol.A 8, 698

Miles, M.J.; Wills, H.W.; The application of STM/AFM to biological molecules; Europ.Microsc.Anal. 6, 7

Miller, J.A.; Hocken, R.J.; Scanning tunneling microscopy bit making on highly oriented pyrolytic graphite: initial results; J.Appl.Phys. 68, 905

Miyasaka, A.; Ogawa, H.; In situ observation of a stinless steel surface in aqueous solutions using scanning tunneling microscope; Corros.Sci. 31, 99

Miyata, C.; Observation of microfabricated structures by STM (in Japanese); J.Soc.Instrum.Contr.Eng. 29, 62

Miyazaki, Y.; Koga, Y., Hayashi, H.; Observation of vickers imprints by scanning tunneling microscopy; J.Vac.Sci.Technol.A 8, 628

Mizutani, W.; Inukai, J., Ono, M.; Making a monolayer hole in a graphite surface by means of a scanning tunneling microscope; Jpn.J.Appl.Phys. 29, L815

Mizutani, W.; Shigeno, M., Ono, M., Kajiumura, K.; Voltage-dependent scanning tunneling microscopy images of liquid crystals on graphite; Appl.Phys.Lett. 56, 1974

Mizutani, W.; Shigeno, M., Sakakibara, Y., Kajimura, K., Ono, M., Tanishima, S., Ohno, K., Toshima, N.; Scanning tunneling spectroscopy study of adsorbed molecules; J.Vac.Sci.Technol.A 8, 675

Möller, R.; Baur, C., Kürz, P., Graf, U.; STM Untersuchungen der GaAs(110)-Oberfläche bei verschiedenen Spannungen; Verh.Dtsch.Phys.Ges. 4, 1157

Möller, R.; Baur, C., Esslinger, A., Graf, U., Kürz, P.; Voltage dependence of the morphology of the GaAs(110) surface observed by scanning tunneling microscopy; Nanotechn. 1, 50

Möller, R.; Baur, C., Graf, U., Kürz, P., Hänsch, T.W.; STM Untersuchung isotroper und anisotroper Silber-Inseln auf ITO-Glas; Verh.Dtsch.Phys.Ges. 4, 1109

Möller, R.; Coenen, R., Esslinger, A., Koslowski, B.; The topography of isolated molecules of copper-phthalocyanine on GaAs(110); J.Vac.Sci.Technol.A 8, 659

Möller, R.; Esslinger, A., Rauscher, M.; Tunneling tips imaged by scanning tunneling microscopy; J.Vac.Sci.Technol.A 8, 434

Möller, R.; Esslinger, A., Koslowski, B.; Thermisches Rauschen des Tunnelstroms: Eine neue Methode der Rastermikroskopie; Verh.Dtsch.Phys.Ges. 4, 1109

Möller, R.; Esslinger, A., Koslowski, B.; Thermal noise in vacuum scanning tunneling microscopy at zero bias voltage; J.Vac.Sci.Technol.A 8, 590

Moiseev. Yu.N.; Mostepanenko, V.M.; Experimental and theoretical study of the forces and spatial resolution in an atomic-force microscope; Sov.Phys.-Tech.Phys. 35, 84

Moreland, J.; Rice, P.; High-resolution, tunneling-stabilized magnetic imaging and recording; Appl.Phys.Lett. 57, 310

Mori, Y.; Yamauchi, K., Endo, K., Ide, T., Toyota, H., Nishizawa, K., Hasegawa, M.; Evaluation of eleastic emission machined surfaces by scanning tunneling microscopy; J.Vac.Sci.Technol.A 8, 621

Moyer, P.J.; Jahncke, C.L., Paesler, M.A., Reddick, R.C., Warmack, R.J.; Spectroscopy in the evanescent field with an analytical photon scanning tunneling microscope; Phys.Lett.A 145, 343

Mo, Y.-W.; Kariotis, R., Swartzentruber, B.S., Webb, M.B., Lagally, M.G.; Scanning tunneling microscopy study of diffusion, growth, and coarsening of Si on Si(001); J.Vac.Sci.Technol.A 8, 201

Murday, J.S.; Colton, R.J.; Proximal probes: techniques for measuring at the nanometer scale; Mater.Sci.Eng.B 6, 77

Murooka, K.-I.; Mitome, M., Tanishiro, Y., Takayanagi, K.; High-resolution electron microscope observation of "atomic bridge" formation between two interacting gold particles; J.Vac.Sci.Technol.A 8, 153

Musselman, I.H.; Peterson, P.A., Russell, P.E.; Fabrication of tips with controlled geometry for scanning tunneling microscopy; Prec.Eng. 12, 3

Musselman, I.H.; Russell, P.E.,; Platinum/Iridium tips with controlled geometry for scanning tunneling microscopy; J.Vac.Sci.Technol.A 8, 3558

Nagahara, L.A.; Thundat, T., Lindsay, S.M.; Nanolithography on semiconductor surfaces under an etching solution; Appl.Phys.Lett. 57, 270

Nagahara, L.A.; Thundat, T., Oden, P.I., Lindsay, S.M., Rill, R.L.; Electrochemical deposition of molecular adsorbates for in situ scanning probe microscopy; Ultramicroscopy 33, 107

Nakagawa, Y.; Ishitani, A., Takahagi, T., Kuroda, H., Tokumoto, H., Ono, M., Kajimura, K.; Scanning tunneling microscopy of silicon surfaces in air: observation of atomic images; J.Vac.Sci.Technol.A 8, 262

Nakagiri, N.; Kaizuka, H.; Simulations of STM images and work function for rough surfaces; Jpn.J.Appl.Phys. 29, 744

Nakamura, T.; Kimura, T., Kimura, T., Ihara, M., Umebu, I.; Scanning tunneling microscopy of chemical vapor deposition grown BiSrCaCuO films; J.Vac.Sci.Technol.A 8, 472

Nejoh, H.; Scanning tunneling microscopy/spectroscopy of a-Si:H/SiNx interface of thin film transistors; J.Vac.Sci.Technol.A 8, 275

Nejoh, H.; Visible mechanism of liquid crystals on graphite under scanning tunneling microscopy; Appl.Phys.Lett. 57, 2907

Neubauer, G.; Cohen, S.R., McClelland, G.M., Horne, D., Mate, C.M.; Force microscopy with bidirectional capacitance sensor; Rev.Sci.Instrum. 61, 2296

Niedermann, Ph.; Tenner, Ch., Kent, A.D., Fischer, O.; Study of field-emitting microstructures using a scanning tunneling microscope; J.Vac.Sci.Technol.A 8, 594

Niehus, H.; Raunau, W., Besocke, K., Spitzl, R., Comsa, G.; Surface structure of NiAl(111) determined by ion scattering and scanning tunneling microscopy; Surf.Sci. 225, L8

Nishikawa, O.; Tomitori, M., Iwawaki, F., Hirano, N.; Correlation between scanning tunneling microscopy/spectroscopy images and apex profiles of scanning tips; J.Vac.Sci.Technol.A 8, 421

Niwa, M.; Iwasaki, H., Hasegawa, S.; Hydrogen terminated Si(100) surfaces studied by scanning tunneling microscopy, x-ray photon spectroscopy, and Auger electron spectroscopy; J.Vac.Sci.Technol.A 8, 266

Nöldechen, A.; Unglaubliche Dichte; HighTech 2, 90

Nomura, K.; Ichimura, K.; Scanning tunneling microscopy observation of sliding charge-density wave in blue bronze; J.Vac.Sci.Technol.A 8, 504

Ogletree, D.F.; Ocal, C., Marchon, B., Somorjai, G.A., Salmeron, M., Beebe, T., Siekhaus, W.; Scanning tunneling microscopy study of the structure of sulfur (23x23)R 30 overlayer on rhenium (0001); J.Vac.Sci.Technol.A 8, 297

Ohnishi, S.; Tsukada, M.; Effect of microscopic electronic states of the tip on scanning tunneling microscopy image; J.Vac.Sci.Technol.A 8, 174

Okumura, A.; Goshi, Y.; A method of scanning tunneling microscope tip approach and several means to improve S/N ratio; Sol.State Phys.(Jpn.) 25, 434

Okumura, A.; Miyamura, K., Goshi, Y.; A scanning tunneling microscopy/current imaging tunneling spectroscopy investigation of an organic molecule: 4-cyano-4'-n-heptylbiphenyl; J.Vac.Sci.Technol.A 8, 625

Oshio, T.; Sakai, T., Ehara, S., Takashima, K.; STM observation of grating surfaces; Mod.Phys.Lett.B 4, 181

Oshio, T.; Tanaka, J., Ono, A., Nagamura, T., Kamizono, Y., Sakai, Y., Ehara, S.; Tunneling spectroscopy of conductive ceramics; J.Vac.Sci.Technol.A 8, 455

Otsuka, I.; Iwasaki, T.; Chloride ion-induced reconstruction on an electrochemically roughened Ag electrode; J.Vac.Sci.Technol.A 8, 530

Packard, W.E.; Ning Dai, Dow, J.D., Jaklevic, R.C., Kaiser, W.J., Tang, S.L.; Externally strained Si(100) observed with scanning tunneling microscopy; J.Vac.Sci.Technol.A 8, 3512

Pagnia, H.; Radojewski, J., Sotnik, N.; Operation conditions of an optical STM; Optik 86, 87

Pagnia, H.; Sotnik, N., Wirth, W.; An STM-STP study on carbon films; Mater.Lett. 9 (oder 8), 151

Pagnia, H.; Sotnik, N., Wirth, W.; Scanning tunneling microscopic investigations of electroformed planar metal-insulator-metal diodes; Int.J.Electron. 69, 25

Pancorbo, M.; Anguiano, E., Diaspro, A., Aguilar, M.; A Wiener filter with circular-aperture-like point spread function to restore scanning tunneling microscopy (STM) images; Patt.Recogn.Lett. 11, 553

Pashley, M.D.; The application of scanning tunneling microscopy to the study of molecular beam epitaxy; J.Cryst.Growth 99, 473

Pashley, M.D.; The application of scanning tunneling microscopy to the study of molecular beam epitaxy; J.Cryst.Growth 99, 473

Peale, D.R.; Cooper, B.H.; A scanning tunneling microscope for ultrahigh vacuum atom-surface interaction studies; J.Vac.Sci.Technol.A 8, 345

Pelz, J.P.; Koch, R.H.; Tip-related artifacts in scanning tunneling potentiometry; Phys.Rev.B 41, 1212

Persch, G.; Engelmann, H., Strack, A.; Untersuchungen an magnetischen Speicherkomponenten mit dem "magnetic force" Mikroskop; Verh.Dtsch.Phys.Ges. 4, 1108

Poirier, G.E.; White, J.M.; Diffraction grating calibration of scanning tunneling microscope piezoscanners; Rev.Sci.Instrum. 61, 3917

Prater, C.B.; Scanning ion-conductance microscope and atomic force microscope; Scanning 12, 50

Prater, C.B.; Butt, H.J., Hansma, P.K.; Atomic force microscopy; Nature 345, 839

Probst, O.; Grafström, S., Kowalski, J., Neumann, R., Wörtge, M.; Ein Kraftmikroskop mit Signalnachweis durch Tunnelstrom und auf einem Trägheitsmechanismus beruhender Tunnelspitzenannäherung; Verh.Dtsch.Phys.Ges. 4, 1106

Qian, J.J.; Wang, Y.T., Ho, J., Hsu, C.C.; Atomic structure at the last LaSi2-x/Si(100) interface; Appl.Phys.Lett. 57, 43

Qin, X.-R.; Kirczenow, G.; Theory of scanning-tunneling-microscopy images of intercalated graphite surfaces; Phys.Rev.B 41, 4976

Rabe, J.P.; Buchholz, St., Ritcey, A.M.; Reactive graphite etch and the structure of an adsorbed organic monolayer - a scanning tunneling microscopy study; J.Vac.Sci.Technol.A 8, 679

Ramos, M.M.D.; Stoneham, A.M., Sutton, A.P., Pethica, J.B.; Effects of the STM tip on atomic positions: an explanation for the nonobservation of adsorbed molecules?; J.Phys., Cond.Matt. 2, 5913

Reiss, G.; Schneider, F., Vancea, J., Hoffman, H.; Scanning tunneling microscopy on rough surfaces: deconvolution of constant current images; Appl.Phys.Lett. 57, 867

Reiss, G.; Vancea, J., Wittmann, H., Zweck, J., Hoffmann, H.; Scanning tunneling microscopy on rough surfaces: tip-shape-limited resolution; J.Appl.Phys. 67, 1156

Renner, Ch.; Niedermann, Ph., Kent, A.D., Fischer, O.; A versatile low-temperature scanning tunneling microscope; J.Vac.Sci.Technol.A 8, 330

Roberts, C.J.; Hoffamnn-Millack, B., Steer, W.S.; Sequential imaging of an amorphous platinum surface down to atomic resolution by STM in air; Meas.Sci.Technol. 1, 199

Robinson, R.S.; Real-time in situ studies of metal electrodeposition with scanning tunneling microscopy; J.Vac.Sci.Technol.A 8, 511

Rohrbeck, W.; Chilla, E., Riedel, J., Fröhlich, H.-J.; Nachweis von akustischen Oberflächenwellen mittels eines Rastertunnelmikroskops; BEDO 23, 319

Rong, Z.Y.; Cohen, L.F., Wolf, E.L.; Scanning tunneling microscope observations of the Coulomb blockade; Phys.Lett.A 146, 281

Rousset, S.; Gauthier, S., Siboulet, O., Sacks, W., Belin, M., Klein, J.; Superstructures of sulfur on stepped copper surfaces observed by scanning tunneling microscopy; J.Vac.Sci.Technol.A 8, 302

Rugar, D.; Hansma, P.; Atomic force microscopy; Physics Today 43, 23

Rugar, D.; Mamin, H.J., Güthner, P., Lambert, S.E., Stern, J.E., McFayden, I., Yogi, T.; Magnetic force microscopy: general principles and application to longitudinal recording media; IBM Res.Rep. RJ 7257 (68024) 1/16/90

Rugar, D.; Mamin, H.J., Güthner, P., Lambert, S.E., Stern, J.E., McFadyen, I., Yogi, T.; Magnetic force microscopy: general principals and application to longitudinal recording media; J.Appl.Phys. 68, 1169

Sakamaki, K.; Itoh, K., Fujishima, A., Goshi, Y.; In situ scanning tunneling microscopy observation of surface roughness of electrochemically modified Ag under potentiostatic control in conjunction with surface-enhanced Raman scattering; J.Vac.Sci.Technol.A 8, 525

Sakamaki, K.; Itoh, K., Fujishima, A., Gohshi, Y.; Surface density of states of TiO2(110) single crystal and adsorbed molecular observation by scanning tunneling microscopy and tunneling spectroscopy; J.Vac.Sci.Technol.A 8, 614

Sakurai, T.; Hashizume, T., Kamiya, I., Hasegawa, Y., Sano, N., Pickering, H.W., Sakai, A.; Field ion-scanning tunneling microscopy; Prog.Surf.Sci. 33, 3

Sakurai, T.; Hashizume, T., Hasegawa, Y., Kamiya, I., Sano, N., Yokoyama, K., Tanaka, H., Sumita, I.; New versatile room-temperature field ion scanning tunneling microscopy; J.Vac.Sci.Technol.A 8, 324

Sakurai, T.; Hasegawa, Y., Hashizume, T., Kamiya, I., Ide, T., Sumita, I., Pickering H.W., Hyodo, S.; Atomic hydrogen chemisorption on the Si(111) 7x7 surface; J.Vac.Sci.Technol.A 8, 259

Salmeron, M.; Beebe, T., Odriozola, J., Wilson, T., Ogletree, D.F., Siekhaus, W.; Imaging of biomolecules with the scanning tunneling microscope: problems and prospects; J.Vac.Sci.Technol.A 8, 635

Sarayeddine, K.; Spajer, M., Courjon, D., Vigoureux, J.M.; Scanning tunneling optical microscopy: a new tool for high resolution non destructive testing; EMAG MICRO 89, Proc.Inst.Phys.Electr.Microsc.Anal. 1, 29

Sarid, D.; Iams, D.A., Ingle, J.T., Weissenberger, V., Ploetz, J.; Performance of a scanning force microscope using a laser diode; J.Vac.Sci.Technol.A 8, 378

Saulys, D.; Rudd, G., Garfunkel, E.; Scanning tunneling microscopy assisted oxide surface etching; Appl.Phys.Lett. (preprint)

Schlüter, G.W.B.; Raster-Sondenmikroskopie - Eine Klasse neuer mikroskopischer Verfahren; Jahrb.Opt.Feinmech. 38, 16 (1991)

Schmidt, H.; Heil, J., Wesner, J., Grill, W.; Atomic force sensors constructed from carbon and quartz fibers; J.Vac.Sci.Technol.A 8, 388

Schönenberger, C.; Alvarado, S.F., Lambert, S.E., Sanders, I.L.; Separation of magnetic and topographic effects in force microscopy; J.Appl.Phys. 67, 7278

Schweizer, E.K.; Eigler, D.M.; Untersuchung von adsorbiertem Xe auf Ni(110) und Pt(111) mit einem 4 K Raster-Tunnelmikroskop; Verh.Dtsch.Phys.Ges. 4, 1157

Selci, S.; Cricenti, A., Felici, A.C., Generosi, R., Gori, E., Djaczenko, W., Chiarotti, G.; Molecular structure of organic compounds observed by high resolution scanning tunneling microscopy; J.Vac.Sci.Technol.A 8, 642

Sexton, B.A.; Cotterill, G.F., Fletcher, St., Horne, M.D.; A scanning tunneling microscopy study of the surface microstructure of alpha- and beta-lead dioxide; J.Vac.Sci.Technol.A 8, 544

Shedd, G.M.; Russell, Ph.E.; The scanning tunneling microscope as a tool for nanofabrication; Nanotechn. 1, 67

Shigeno, M.; Mizutani, W., Suginoya, M., Ohmi, M., Kajimura, K.; Observation of liquid crystals on graphite by scanning tunneling microscopy; Jpn.J.Appl.Phys. 29, L119

Shih, C.K.; Feenstra, R.M., Martensson, P.; Scanning tunneling microscopy and spectroscopy of thin metal films on the GaAs(110) surface; J.Vac.Sci.Technol.A 8, 3379

Shimizu, N.; Kimura, T., Nakamura, T., Umebu, I.; An ultrahigh vacuum scanning tunneling microscope with a new inchworm mechanism; J.Vac.Sci.Technol.A 8, 333

Slough, C.G.; Giambattista, B., McNairy, W.W., Coleman, R.V.; Scanning tunneling microscopy of the linear chain compounds NbSe3, and TaSe3; J.Vac.Sci.Technol.A 8, 490

Smith, I.; Howland, R.; Applications of scanning probe microscopy in the semiconductor industry; Sol.State Technol. Dec., 53

Snellman, H.; Pelliniemi, L.J., Penttinen, R., Laiho, R.; Application of scanning tunneling microscopy for imaging of CNBr-peptides of type I collagen; J.Vac.Sci.Technol.A 8, 692

Soto, M.R.; The efeect of point defects on the STM image of graphite; Surf.Sci. 225, 190

Spence, J.C.H.; Lo, W., Kuwabara, M.; Observation of the graphite surface by reflection electron microscopy during STM operation; Ultramicroscopy 33, 69

Steer, W.S.; et al.; A versatile STM for use in air; Meas.Sci.Technol. 1, 881

Strecker, H.; Krameyer, C., Stahl, C.; Materialkontrast mit dem Rastertunnelmikroskop bei Betrieb an Luft; Verh.Dtsch.Phys.Ges. 4, 1109

Strecker, H.; Persch, G.; Application of STM in magnetic storage device manufacturing; J.Vac.Sci.Technol.A (1991) (preprint)

Strecker, H.; Stahl, C., Starke, H.; Material contrast using the scanning tunneling microscope in ambient conditions; J.Vac.Sci.Technol.A 8, 618

Stroscio, J.A.; First, P.N., Dragoset, R.A., Whitman, L.J., Pierce, D.T., Celotta, R.J.; Dispersion of evanescent band gap states in Fe clusters on GaAs(110); J.Vac.Sci.Technol.A 8, 284

Su Ya-Xian; Jiao Yue-Kan, Xu San-Duo, Yao Jun-En, Lin Ke-Chun; The surface structure of artificial and natural membranes as studied by scanning tunneling microscopy; J.Vac.Sci.Technol.A 8, 695

Sugawara, Y.; Ishizaka, T., Morita, S., Imai, S., Mikoshiba, N.; Simultaneous observation of atomically resolved AFM/STM images of a graphite surface; Jpn.J.Appl.Phys. 29, L296

Sugawara, Y.; Ishizaka, T., Morita, S.; Origin of anomalous corrugation hight of STM images of graphite; Jpn.J.Appl.Phys. 29, 1533

Sugawara, Y.; Ishizaka, T., Morita, S.; Simulraneous imaging of a graphite surface with atomic force / scanning tunneling microscope (AFM/STM); Jpn.J.Appl.Phys. 29, 1539

Sugawara, Y.; Ishizaka, T., Morita, S., Imai, S., Mikoshiba, N.; Anomalous corrugation height of atomically resolved AFM images of a graphite surface; Jpn.J.Appl.Phys. 29, L502

Sugihara, K.; Sakai, A., Akama, Y., Shoda, N., Kato, Y., Tokumoto, H., Murakami, H.; An ultrahigh vacuum tip transfer system for the scanning tunneling microscope; Rev.Sci.Instrum. 61, 81

Sumetskii, M.; STM imaging of a nanometer conducting structure in the exponential approximation; J.Phys., Cond.Matt. 2, 2487

Sumetskii, M.; STM imaging of a nanometer conducting structure in the exponential approximation; J.Phys., Cond.Matt. 2, 2487

Sumita, I.; Yokotsuka, T., Tanaka, H., Udagawa, M., Watanabe, Y., Takao, M., Yokoyama, K.; Observation of domain boundaries on the Si(111)7x7 surface by scanning tunneling microscope; Appl.Phys.Lett. 57, 1313

Suzuki, M.; Maruno, T., Yamamoto, F., Nagai, K.; Surface roughness of rubbed polyimide film for liquid crystals by scanning tunneling microscopy; J.Vac.Sci.Technol.A 8, 631

Swartzentruber, B.S.; Mo, Y.-W., Webb, M.B., Lagally, M.G.; Observations of strain effects on the Si(001) surface using scanning tunneling microscopy; J.Vac.Sci.Technol.A 8, 210

Takeuchi, K.; Uehara, Y., Ushioda, S., Mikoshiba, N., Morita S.; Power spectra of surface roughness of light emitting tunnel junctions measured by scanning tunneling microscopy; J.Vac.Sci.Technol.A 8, 557

Tamura, A.; Statistical treatment on the formation of Si(100)2xn surface; J.Vac.Sci.Technol.A 8, 192

Tanaka, I.; Kato, T., Ohkouchi, S., Osaka, F.; Observation of multiquantum well structure in air using a scanning tunneling microscope; J.Vac.Sci.Technol.A 8, 567

Tanaka, M.; Yamazaki, S., Fujinami, M., Takahashi, T., Katayama-Yoshida, H., Mizutani, W., Kajimura, K., Ono, M.; Surface electronic properties on BiO plane pf Bi2Sr2CaCu2O8 single crystal measured by scanning tunneling microscopy; J.Vac.Sci.Technol.A 8, 475

Tanimoto, M.; Nakano, Y.; Observation of double heterostructures for laser diodes using scanning tunneling microscopy and current imaging tunneling spectroscopy in air; J.Vac.Sci.Technol.A 8, 553

Tear, S.P.; The use of STM for surface structure determination; Europ.Microsc.Anal. 9, 13

Terashima, K.; Kondoh, M., Yoshida, T.; Fabrication of nucleation sites for nanometer size selective deposition by scanning tunneling microscope; J.Vac.Sci.Technol.A 8, 581

Terris, B.D.; Stern, J.E., Rugar, D., Mamin, H.J.; Localized charge force microscopy; J.Vac.Sci.Technol.A 8, 374

Tersoff, J.; Role of tip electronic structure in scanning tunneling microscope images; Phys.Rev.B 41, 1235

Tersoff, J.; Lang, N.D.; Tip-dependent corrugation of graphite in scanning tunneling microscopy; Phys.Rev.Lett. 65, 1132

Thundat, T.; Nagahara, L.A., Oden, P., Lindsay, S.M.; Direct observation of bioelectrochemical processes by scanning tunneling microscopy; J.Vac.Sci.Technol.A 8, 645

Thundat, T.; Nagahara, L.A., Oden, I.P., Lindsay, S.M., George, M.A., Glausinger, W.S.; Modification of tantalum surfaces by scanning tunneling microscopy in an electrochemical cell; J.Vac.Sci.Technol.A 8, 3537

Thundat, T.; Nagahara, L.A., Lindsay, S.M.; Scanning tunneling microscopy studies of semiconductor electrochemistry; J.Vac.Sci.Technol.A 8, 539

Tiedje, T.; Brown, A.; Performance limits for scanning tunneling microscope; J.Appl.Phys. 68, 649

Tokumoto, H.; Miki, K., Kajimura, K.; STM images of anisotropic atomic steps on Si(111)-7x7 surfaces; J.Cryst.Growth 99, 1329

Tokumoto, H.; Miki, K., Murakami, H., bando, H., Ono, M., Kajimura, K.; Real-time observation of oxygen and hydrogen adsorption on silicon surfaces by scanning tunneling microscopy; J.Vac.Sci.Technol.A 8, 255

Tokumoto, H.; Okayama, S., Miki, K., Murakami, H., Okano, M., Sakamoto, K., Sakamoto, T., Kawanami, H., Suzuki, E., Wakiyama, S., Ando, K., Ohmira, K., Watanabe, K., Honma, A., Ishihara, K., Inoue, A., Sakai, F., Kobayashi, Y., et al.; UHV-STM system for investigation of compound semiconductors (in Japanese); Bull.Electrotech.Lab. 54, 16

Tokumoto, H.; Wakiyama, S., Miki, K., Okayama, S.; Scanning tunneling microscopy of (11-2) oriented steps on a cleaved Si(111) surface; Appl.Phys.Lett. 56, 743

Tomita, E.; Matsuda, N., Itaya, K.; Surface electronic structure of semiconductor (p- and n-Si) electrodes in electrolyte solution; J.Vac.Sci.Technol.A 8, 534

Tomitori, M.; Iwawaki, F., Hirano, N., Katsuki, F., Nishikawa, O.; Corrugation of Si surfaces and profiles of tip apexes; J.Vac.Sci.Technol.A 8, 222

Tomitori, M.; Hirano, N., Iwawaki, F., Watanabe, Y., Takayanagi, T., Nishikawa, O.; Elaboration and evaluation of tip manipulation of scanning tunneling microscopy; J.Vac.Sci.Technol.A 8, 425

Tonomura, A.; Electron interferometry using a coherent beam from point electron source; J.Vac.Sci.Technol.A 8, 155

Trafas, B.M.; Hill, D.M., Siefert, R.L., Weaver, J.H.; Adsorption and interaction of Sm on GaAs(110) studied by scanning tunneling microscopy; Phys.Rev.B 42, 3231

Troyanovskii, A.M.; Hietschold, M., Khaikin, M.S., Androsch, I., Vollmann, W., Starke, M.; Scanning tunneling microscopy invstigations of organic folm topography; Thin Solid Films 188, 329

Tsukada, M.; Mahanty, J.; A mechanism for Coulomb blockade in scanning tunneling spectroscopy; Phys.Lett.A 148, 367

Tsukada, M.; Kobayashi, K., Ohnishi, S.; First-principles theory of the scanning tunneling microscopy simulation; J.Vac.Sci.Technol.A 8, 160

Uosaki, K.; Kita, H.; In situ, real-time monitoring of electrode surfaces by scanning tunneling microscopy: III, Surface structure of Pt and Pd electrodes; J.Vac.Sci.Technol.A 8, 520

Utsugi, Y.; Nanometre-scale chemical modification using a scanning tunneling microscope; Nature 347, 747

Valdés, J.; Choleva, E.· Possible evidence of single electron tunneling with the scanning tunneling microscope at room temperature; J.Vac.Sci.Technol.A 8, 598

Vinogradova, S.A.; Golubok, A.O., Kolomytkin, O.V., Tipisev, S.Ya.; Scanning tunneling microscopy of gramicidin A molecules embedded in Langmuir-Blodgett film; JETP Lett. 51, 581

Völcker, M.; Krieger, W., Suzuki, T., Thost, J.-P., Walther, H.; Messungen mit einem Laser-Rastertunnelmikroskop; Verh.Dtsch.Phys.Ges. 4, 1159

Wadas, A.; Güntherodt, H.-J.; Lateral resolution in magnetic force microscopy. Application to periodic structures; Phys.Lett.A 146, 277

Wadas, A.; Grutter, P., Güntherodt, H.-J.; Analysis of in-plane bit structure by magnetic force microscopy; J.Appl.Phys. 67, 3462

Wadas, A.; Grütter, P., Güntherodt, H.-J.; Analysis of magnetic bit patterns by magnetic force microscopy; J.Vac.Sci.Technol.A 8, 416

Walls, M.G.; Tip-generated artefacts and their avoidance in STM; Proc.Roy.Microsc.Soc. 25, S59

Watanabe, M.O.; Tanaka, K., Sakai, A.; High-temperature scanning tunneling microscope; J.Vac.Sci.Technol.A 8, 327

Weisenhorn, A.L.; Mac Dougall, J.E., Gould, S.A.C., Coc, S.D., Wise, W.S., Massie, J., Maivald, P., Elings, V.B., Stucky, G.D., Hansma, P.K.; Imaging and manipulating molecules on a zeolite surface with an atomic force microscope; Science 247, 1330

Wengelnik, H.; Neddermeyer, H.; Oxygen-indiced sharpening process of W(111) tips for scanning tunneling microscope use; J.Vac.Sci.Technol.A 8, 438

Wichmann, B.; Eerden, J.P. van der, Gerritsen, J.W.; In situ STM of silver electrocrystallization; J.Cryst.Growth 99, 1333

Wickramasinghe, H.K.; Scanning probe microscopy: current status and future trends; J.Vac.Sci.Technol.A 8, 363

Wiesendanger, R.; Local structures of the Si(100) surface studied by scanning tunneling microscopy; Surf.Sci. 232, 1

Wiesendanger, R.; Anselmetti, D., Güntherodt, H.-J.; Advances in STM design and instrumentation; Europhys.News 21, 72

Wiesendanger, R.; Scanning tunneling microscopy on laser- and thermal-annealed Si(111): transitions from 7x7 reconstructed to disordered surface structures; Ultramicroscopy 32, 291

Wiesendanger, R.; Buergler, D., Tarrach, G., Anselmetti, D., Hidber, H.R., Güntherodt, H.-J.; An ultrahigh vacuum scanning tunneling microscope for the investigation of clean surfaces; J.Vac.Sci.Technol.A 8, 339

Wiesendanger, R.; Güntherodt, H.-J.; Observation of vacuum tunneling of spin-polarized electrons with the scanning tunneling microscope; Phys.Rev.Lett. 65, 247

Wiesendanger, R.; Güntherodt, H.-J.; Surface step and defect structure of Cr(001) studied by scanning tunneling microscopy; Surf.Sci. 235, 1

Wiesendanger, R.; Güntherodt, H.-J.; Observation of vacuum tunneling of spin-polarized electrons with the scanning tunneling microscope; Phys.Rev.Lett. 65, 247

Wiesendanger, R.; Güntherodt, H.-J.; Surface step and defect structure of Cr(001) studied by scanning tunneling microscopy; Surf.Sci. 235, 1

Wiesendanger, R.; Güntherodt, H.-J., Güntherodt, G., Gambino, R.J., Ruf, R.; Scanning tunneling microscopy with spin-polarized electrons; Z.Phys.B 80, 5

Wiesendanger, R.; Tarrach, G., Burgler, D., Güntherodt, H.-J.; Scanning tunneling microscopy study of Si(111)7x7 in the presence of multi-step edges; Europhys.Lett. 12, 57

Wilkins, R.; Amman, M., Soltis, R.E., Ben-Jacob, E., Jaklevic, R.C.; Tunneling spectroscopy study of YBa2Cu3O7 thin films using a cryogenic scanning tunneling microscope; Phys.Rev.B 41, 8904

Williams, C.C.; Wickramasinghe H.K.; Microscopy of chemical-potential variations on a atomic scale; Nature 344, 317

Wilson, R.J.; Meijer, G., Bethune, D.S., Johnson, R.D., Chambliss, D.D., Vries, M.S. de, Hunziker, H.E., Wendt, H.R.; Imaging C60 clusters on a surface using a scanning tunneling microscope; Nature 348, 621

Worcester, D.L.; Kim, H.S., Miller, R.G., Bryant, P.J.; Imaging bacteriorhodopsin lattices in purple membranes with atomic force microscopy; J.Vac.Sci.Technol.A 8, 403

Wormeester, H.; Midden, H.J.P. van, Kip, G.A.M., Sasse, A.G.M.; The use of splines to analyze scanning tunneling microscopy data; J.Vac.Sci.Technol.A 8, 3548

Wragg, J.L.; Chamberlain, J.E., White, H.W., Krätschmer, W., Huffman, D.R.; Scanning tunneling microscopy of solid C60/C70; Nature 348, 623

Xie, Y.; Xu, Q., Tang, Y., Bai, C., Dai, C.; Scanning tunneling microscopy investigation of a spontaneous monolayer dispersion system: HgCl2 on highly oriented pyrolytic graphite; J.Vac.Sci.Technol.A 8, 610

Yagi, A.; Tsukuda, S., Takahashi, Y., Morita, S., Sugawara, Y., Okada, T., Imai, S., Mikoshiba, N.; Differential conductance imaging under alternate current tunneling bias; J.Vac.Sci.Technol.A 8, 336

Yang, R.; Evans, D.F., Christensen, L., Hendrickson, W.A.; Scanning tunneling microscopy evidence of semicrystalline and helical conducting polymer structures; J.Phys.Chem. 94, 6117

Yang, R.; Yang, X.R., Evans, D.F., Hendrickson, W.A., Baker, J.; Scanning tunneling microscopy images of poly(ethylene oxide) polymers: evidence for helical and superhelical structures; J.Phys.Chem. 94, 6123

Yao, J.E.; Jiao, Y.K.; Observing the varying scanning tunneling microscopy image of graphite; J.Vac.Sci.Technol.A 8, 508

Yasutake, M.; Miyata, C.; Scanning tunneling microscope combined with optical microscope for large sample measurement; J.Vac.Sci.Technol.A 8, 350

Yau, S.-T.; Saltz, D., Nayfeh, M.H.; Laser-assisted deposition of nanometer structures using a scanning tunneling microscope; Appl.Phys.Lett. 57, 2913

Yokohata, T.; Kato, K., Ohmura, K.; Nanometer-scale roughness study and indentation test with a scanning tunneling microscope; J.Vac.Sci.Technol.A 8, 585

Yoshida, S.; Iizuka, K.; Approaches to nanometer technology - a review of the Yoshida Nano-Mechanism Project; Nanotechn. 1, 13

Yoshimura, M.; Fujita, K., Ara, N., Kageshima, M., Shioda, R., Kawazu, A., Shigekawa, H., Hyodo, S.; Observation of an organic superconductor [bis(ethylenedithio)tetrathiafulvalene]2[Cu(NCS)2] by scanning tunneling microscopy; J.Vac.Sci.Technol.A 8, 488

Youngquist, M.G.; Driscoll, R.J., Coley, T.R., Goddard, W.A., Baldeschwieler, J.D.; Scanning tunneling microscopy of DNA: atom-resolved imaging, general observations and possible contrast mechanism; J.Vac.Sci.Techn.A (1991) (preprint)

Zeglinski, D.M.; Ogletree, D.F., Beebe Jr., T.P., Hwang, R.Q., Somorjai, G.A., Salmeron, M.B.; An ultrahigh vacuum scanning tunneling microscope for surface science studies; Rev.Sci.Instrum. 61, 3769

Zhang, X.G.; Stimming, U.; Scanning tunneling microscopy of copper corrosion in aqueous percloric acid; Corros.Sci. 30, 951

Zheng, N.J.; Tsong, I.S.T.; Rsonant-tunneling theory of imaging close-packed metal surfaces by scanning tunneling microscopy; Phys.Rev.B 41, 2671

ESPRIT Basic Research Series

J. W. Lloyd (Ed.): **Computational Logic.** Symposium Proceedings, Brussels, November 1990. XI, 211 pages. 1990

E. Klein, F. Veltman (Eds.): **Natural Language and Speech.** Symposium Proceedings, Brussels, November 1991. VIII, 192 pages. 1991

G. Gambosi, M. Scholl, H.-W. Six (Eds.): **Geographic Database Management Systems.** Workshop Proceedings, Capri, May 1991. VIII, 320 pages. 1992

R. Kassing (Ed.): **Scanning Microscopy.** Symposium Proceedings, Wetzlar, October 1990. X, 207 pages. 1992

MIX
Papier aus verantwortungsvollen Quellen
Paper from responsible sources
FSC® C105338

If you have any concerns about our products,
you can contact us on
ProductSafety@springernature.com

In case Publisher is established outside the EU,
the EU authorized representative is:
Springer Nature Customer Service Center GmbH
Europaplatz 3, 69115 Heidelberg, Germany

Printed by Libri Plureos GmbH
in Hamburg, Germany